PRACTICAL TREATISE

ON THE

DIFFERENTIAL AND INTEGRAL

CALCULUS,

WITH SOME OF ITS

APPLICATIONS TO MECHANICS AND ASTRONOMY.

BY

WILLIAM G. PECK, PH.D., LL.D.,

Professor of Mathematics and Astronomy in Columbia College and of Mechanics in the School of Mines.

A. S. BARNES & COMPANY,
NEW YORK AND CHICAGO.

PUBLISHERS' NOTICE.

PECK'S MATHEMATICAL SERIES.

CONCISE, CONSECUTIVE, AND COMPLETE.

I. FIRST LESSONS IN NUMBERS.
II. MANUAL OF PRACTICAL ARITHMETIC.
III. COMPLETE ARITHMETIC.
IV. MANUAL OF ALGEBRA.
V. MANUAL OF GEOMETRY.
VI. TREATISE ON ANALYTICAL GEOMETRY.
VII. DIFFERENTIAL AND INTEGRAL CALCULUS.
VIII. ELEMENTARY MECHANICS (without the Calculus).
XI. ELEMENTS OF MECHANICS (with the Calculus).

NOTE.—Teachers and others, discovering errors in any of the above works, will confer a favor by communicating them to us.

PREFACE.

THE following pages are designed to embody the course of "Calculus and its Applications," as now taught in Columbia College. This course being purely optional, is selected by those only who have a taste for mathematical studies, and it is pursued by them more with reference to its utility than as a means of mental discipline. These circumstances have given to the present work a *practical* character that can hardly fail to commend it to those who study the Calculus for the advantages it gives them in the solution of scientific problems. To meet the wants of this class of students, great care has been taken to avoid superfluous matter; the definitions have been revised and abbreviated; the demonstrations have been simplified and condensed; the rules and principles have been illustrated and enforced by numerous examples, so chosen as to familiarize the student with the use of radical and transcendental quantities; and finally, the manner of applying the Calculus has been exemplified by the solution of a variety of problems in Mechanics and Astronomy.

The method employed in developing the principles of the science is essentially that of Leibnitz. This method, generally known as the *method of infinitesimals*, has been adopted for several reasons: *first*, it is the method adopted in all practical investigations; *second*, it is the method most easily explained and most readily comprehended; and *third*, it is a method, as will be shown in the final note, identical in results with the more commonly adopted *method of limits*, differing from it chiefly in its phraseology and in the simplicity of its results.

The author cannot conclude this prefatory note without acknowledging his obligations to those students who, from year to year, have been willing to turn aside from the attractive pursuit of classical learning to engage in the sterner study of those processes that have contributed so much to the progress of modern science. Without their interest and co-operation this book would never have been written. He would also take this opportunity to express his thanks to his distinguished colleague, Professor J. H. VAN AMRINGE, not only for many valuable suggestions made during the progress of the work, but also for much effective labor in reading and correcting the proofs as they came from the press.

COLUMBIA COLLEGE,
November 24th, 1870.

CONTENTS.

PART I.—DIFFERENTIAL CALCULUS.

I. Definitions and Introductory Remarks.

II. Differentiation of Algebraic Functions.

III. Differentiation of Transcendental Functions.

IV. Successive Differentiation and Development of Functions.

V. Differentiation of Functions of two Variables and of Implicit Functions.

PART II.—APPLICATIONS OF THE DIFFERENTIAL CALCULUS.

I. Tangents and Asymptotes.

II. Curvature.

III. Singular Points of Curves.

IV. Maxima and Minima.

V. Singular Values of Functions.

VI. Elements of Geometrical Magnitudes.

VII. Application to Polar Co-ordinates.

VIII. Transcendental Curves.

PART III.—INTEGRAL CALCULUS.

PART IV.—APPLICATIONS OF THE INTEGRAL CALCULUS.

I. Lengths of Plane Curves.

II. Areas of Plane Curves.

III. Areas of Surfaces of Revolution.

IV. Volumes of Solids of Revolution.

PART V.—APPLICATIONS TO MECHANICS AND ASTRONOMY.

I. Centre of Gravity.

II. Moment of Inertia.

III. Motion of a Material Point.

PART I.

DIFFERENTIAL CALCULUS.

I. Definitions and Introductory Remarks.

Classification of Quantities.

1. The quantities considered in Calculus are of two kinds: *constants*, which retain a fixed value throughout the same discussion, and *variables*, which admit of all possible values that will satisfy the equations into which they enter. The former are usually denoted by leading letters of the alphabet, as a, b, c, etc.; and the latter by final letters, as x, y, z, etc.; particular values of variable quantities are denoted by writing them with one or more dashes, as x', y'', z''', etc.

Functions of one or more Variables.

2. Relations between variables are expressed by equations. In an equation between two variables, values may be assigned to one at pleasure; the resulting equation determines the corresponding values of the other. The one to which arbitrary values are assigned is called the *independent variable*, and the remaining one is said to be a *function* of the former. If an equation contain more than two variables, all but one are *independent*, and that one is a *function* of all the others. The fact that a quantity depends on one or more variables may be expressed as follows:

$$y = f(x)\,;\ z = \varphi(x, y)\,;\ F(x, y, z) = 0.$$

The *first* shows that *y is a function of x*, the *second* that *z is a function of x and y*, and the *third* that x, y, and z, depend on each other, without pointing out which is a function of the other two.

Geometrical representation of a Function.

3. Every function of one variable may be represented by the ordinate of a curve, of which the variable is the corresponding abscissa. For, let y be a function of x, and suppose x to increase by insensible gradations from $-\infty$ to $+\infty$. For each value of x there will be one or more values of y, and these, if real, will determine the position of a point with respect to two rectangular axes. These points make up a curve, at every point of which the relation between the ordinate and abscissa is the same as that between the function and independent variable. This curve is called *the curve of the function*.

For values of x that give imaginary values of y there are no points; for those that give more than one real value of y there is a corresponding number of points.

In a similar manner it may be shown that a function of two variables represents the ordinate of a surface of which the variables are corresponding abscissas.

Differentials and Differentiation.

4. Of two quantities, that is the *less* whose value is nearer to $-\infty$, and that is the *greater* whose value is nearer to $+\infty$. A quantity is said to *increase* when it approaches $+\infty$, and to *decrease* when it approaches $-\infty$. The ordinates of a curve originate from the axis of X and of any two, that is the greater whose extremity is nearer $+\infty$, and that is the less whose extremity is nearer $-\infty$; in like manner of two abscissas, that is the greater whose extremity is nearer $+\infty$, and that the less whose extremity is nearer $-\infty$.

In what follows we shall suppose the independent variable to increase by the continued addition of a *constant* but *infinitely small* increment. For every change in the value of the variable there is a corresponding change in the value of the function. In some cases, as the variable increases, the function increases; it is then said to be an *increasing function:* in other cases the function decreases as the variable increases; it is then said to be a *decreasing function.* In all cases, the *change in value* is called an *increment;* for increasing functions the increment is *positive,* and for decreasing functions it is *negative.* The increment of the function is always infinitely small, but it is not constant, except in particular cases.

The infinitely small increment of the independent variable is called the *differential of the variable,* and the corresponding *increment* of the function is called the *differential of the function.* Hence, *the differential of a quantity is the difference between two consecutive values of that quantity.* It is to be observed that the difference is always found by taking the first value from the second.

The operation of finding a differential is called *differentiation.* The object of the *differential calculus* is to explain the methods of differentiating functions.

Geometrical Illustration.

5. Let KL be a curve in the plane of the rectangular axes OX and OY, and let OA and OB be two abscissas differing from each other by an infinitely small quantity AB. Through A and B draw ordinates to the curve, and let PR be parallel to OX. OA and OB are *consecutive* abscissas, AP and BQ are *consecutive* ordinates, and P and Q are *consecutive* points of the curve. The part of the curve PQ does not differ sensibly from a straight line, and if it be prolonged toward T, the line PT is tangent to the curve at P. If we denote any abscissa OA by x, and the corresponding ordinate by y, we have,

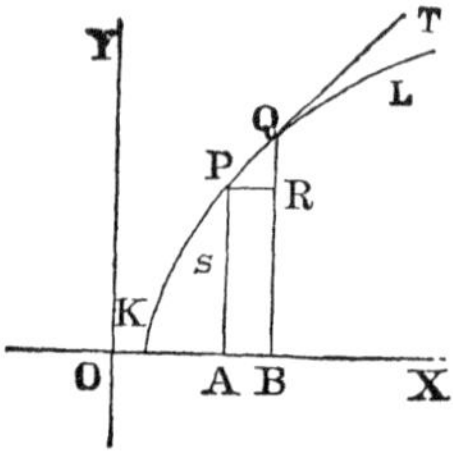

Fig. 1.

$$y = f(x).$$

The line AB is the differential of the independent variable, denoted by the symbol dx; RQ is the differential of the function, denoted by dy; and PQ is the differential of the curve KL, denoted by ds.

The right angled triangle, RPQ, gives the relation $ds = \sqrt{dx^2 + dy^2}$. Denoting the angle RPQ by θ, we have, from trigonometry,

$$\tan\theta = \frac{dy}{dx}; \quad \sin\theta = \frac{dy}{ds} = \frac{dy}{\sqrt{dx^2 + dy^2}};$$

$$\text{and,} \quad \cos\theta = \frac{dx}{ds} = \frac{dx}{\sqrt{dx^2 + dy^2}}.$$

Infinites and Infinitesimals.

6. A quantity is *infinitely great* with respect to another, when the quotient of the former by the latter is greater than any assignable number, and *infinitely small* with respect to it, when the quotient is less than any assignable number. If the term of comparison is *finite*, quantities of the former class are called *infinites*, and those of the latter *infinitesimals*.

Infinites and infinitesimals are of different orders. Let us assume the series,

$$\ldots \frac{a}{x^3}, \frac{a}{x^2}, \frac{a}{x}, a, ax, ax^2, ax^3 \ldots$$

In which a is a finite constant and x variable. If we suppose x to increase, the terms preceding a will diminish, and those following it will increase; when x becomes greater than any assignable quantity, $\frac{a}{x}$ becomes infinitely small *with respect to a*, and because each term bears the same relation to the one that follows it, every term in the series is infinitely small with respect to the following one, and infinitely great with respect to the preceding one. The quantity ax being *infinitely great* with respect to a finite quantity, is called an infinite of the *first* order; ax^2, ax^3, etc., are infinites of the *second, third*, etc., orders. The quantity $\frac{a}{x}$ being infinitely small with respect to a finite quantity is called an infinitesimal of the *first* order; $\frac{a}{x^2}$, $\frac{a}{x^3}$, etc, are infinitesimals of the *second, third*, etc., orders.

It is to be observed that the product of two infinitesimals of the *first* order, is an infinitesimal of the second order. For, let x and y be infinitely small with respect to 1, we shall have,

$$1 : x :: y : xy.$$

Hence, xy bears the same relation to y that x does to 1, that is, it is infinitely small with respect to an infinitesimal of the first order; it is therefore an infinitesimal of the second order. The product of three infinitesimals of the first order is an infinitesimal of the third order, and so on. In general, the product of an infinitesimal of the mth order by one of the nth order, is an infinitesimal of the $(m + n)^{\text{th}}$ order. The product of a finite quantity by an infinitesimal of the nth order, is an infinitesimal of the nth order.

From the nature of an infinite quantity, its value will not be sensibly changed by the addition or subtraction of a finite quantity. A finite quantity may therefore be disregarded in comparison with an infinite quantity. For a like reason an infinitesimal may be disregarded in comparison with a finite quantity, or with an infinitesimal of a lower order. Hence, whenever an infinitesimal is connected, by the sign of addition, or subtraction, with a finite quantity, or with an infinitesimal of a lower order, it may be suppressed without affecting the value of the expression into which it enters.

General method of Differentiation.

7. In order to find the differential of a function, we give to the independent variable its infinitely small increment, and find the corresponding value of the function; from this we subtract the preceding value and reduce the result to its simplest form; we then suppress all infinitesimals which are added to, or subtracted from, those of a lower order, and the result is the differential required.

This method of proceeding is too long for general use, and is only employed in deducing rules for differentiation.

II. Differentiation of Algebraic Functions.

Definition of an Algebraic Function.

8. An *algebraic* function is one in which the relation between the function and its variable can be expressed by the *ordinary* operations of algebra, that is, *addition, subtraction, multiplication, division, formation of powers denoted by constant exponents*, and *extraction of roots indicated by constant indices*. Thus,

$$y^2 = 2px, y = ax^2 - \sqrt{bx}, \text{ and } \sqrt{y} = \sqrt[3]{a^2x - bx^2},$$

are algebraic functions.

Differential of a Polynomial.

9. Let a and c be constants, and r, s, t, functions of x; assume

$$y = ar + s - t + c; \quad . \quad . \quad . \quad . \quad . \quad (1)$$

in which y denotes the polynomial in the second member, and is therefore a function of x.

If we give to x the increment dx, the functions y, r, s, and t, receive corresponding increments dy, dr, ds, dt, and we have,

$$y + dy = a(r + dr) + (s + ds) - (t + dt) + c \quad . \quad . \quad . \quad . \quad . \quad (2)$$

subtracting (1) from (2), we have,

$$dy = adr + ds - dt \quad . \quad . \quad . \quad . \quad . \quad (3)$$

Hence, to differentiate a polynomial, *differentiate each term separately and take the algebraic sum of the results.*

Comparing (1) and (3) we see, *first*, that a *constant factor* remains unchanged, and *secondly*, that a *constant term* disappears by differentiation.

Differential of a Product.

10. Let r and s be functions of x. Placing their product equal to y, we have,

$$y = rs \quad . \; . \; . \; . \; . \quad (1)$$

Giving to x the increment dx, we have, as before,

$$y + dy = (r + dr)(s + ds) = rs + rds + sdr + drds. \; . (2)$$

Subtracting (1) from (2) and suppressing $drds$, which is an infinitesimal of the second order, we have, after replacing y by its value rs,

$$d(rs) = rds + sdr \quad . \; . \; . \; . \; . \quad (3)$$

Hence, to differentiate the product of two functions, *multiply each by the differential of the other, and take the algebraic sum of the results.*

If we suppose $r = tw$, we have, from the rule,

$$dr = tdw + wdt;$$

which substituted in (3), gives,

$$d(stw) = twds + stdw + swdt \quad . \; . \; . \; . \; . \quad (4)$$

In like manner the principle may be extended to the product of any number of functions. Hence, to differentiate the product of any number of functions, *multiply the differential of each by the continued product of all the others, and take the algebraic sum of the results.*

COR.—If we divide both members of (4) by stw, we have,

$$\frac{d(stw)}{stw} = \frac{ds}{s} + \frac{dt}{t} + \frac{dw}{w} \quad . \; . \; . \; . \; . \quad (5)$$

and similarly, where there are a greater number of factors. Hence, *the differential of a product divided by that product, is equal to the sum of the quotients obtained by dividing the differential of each factor by that factor.*

Differential of a Fraction.

11. Let s and t be functions of x, and assume,

$$y = \frac{s}{t} \quad . \; . \; . \; . \; . \quad (1)$$

Giving to x the increment dx, we have,

$$y + dy = \frac{s + ds}{t + dt} \quad . \; . \; . \; . \; . \quad (2)$$

Subtracting (1) from (2),

$$dy = \frac{s + ds}{t + dt} - \frac{s}{t} = \frac{tds - sdt}{t^2 + tdt}.$$

Replacing y by its value $\frac{s}{t}$, and suppressing tdt in comparison with t^2, we have,

$$d\left(\frac{s}{t}\right) = \frac{tds - sdt}{t^2} \quad . \; . \; . \; . \; . \quad (3)$$

Hence, the differential of a fraction is equal to *the denominator into the differential of the numerator, minus the numerator into the differential of the denominator, divided by the square of the denominator.*

If either term of the fraction be constant, its differential will be 0. When the denominator is constant, formula (3) becomes,

$$d\left(\frac{s}{c}\right) = \frac{ds}{c}; \quad . \; . \; . \; . \; . \quad (4)$$

when the numerator is constant it becomes,

$$d\left(\frac{c}{t}\right) = -\frac{cdt}{t^2} \quad . \; . \; . \; . \; . \quad (5)$$

Differential of a Power.

12. Let s be a function of x, and m a constant. Assume

$$y = s^m \quad \ldots\ldots \quad (1)$$

Giving to x the increment dx, we have,

$$y + dy = (s + ds)^m = s^m + ms^{m-1}ds$$

$$+ \frac{m(m-1)}{1.2} s^{m-2}ds^2 + (\text{etc.}) \quad \ldots\ldots \quad (2)$$

In which all the terms of the development, after the second, contain the square, or some higher power, of ds. Subtracting (1) from (2), we have,

$$dy = ms^{m-1}ds + (\text{etc.})\, ds^2.$$

Suppressing all the terms of the second member, after the first, in accordance with the principle laid down in Art. 6, and replacing y by its value, we have,

$$d(s^m) = ms^{m-1}ds \quad \ldots\ldots \quad (3)$$

Hence, to differentiate any power of a function, *diminish the exponent of the function by* 1, *and multiply the result by the primitive exponent and the differential of the function.*

Differential of a Radical.

13. Let s be a function of x, and assume,

$$y = \sqrt[n]{s} \quad \ldots\ldots \quad (1)$$

We may write (1) in the form,

$$y = s^{\frac{1}{n}} \quad . \quad . \quad . \quad . \quad . \quad (2)$$

Differentiating (2), as in the last article, we have,

$$dy = \frac{1}{n} s^{\frac{1}{n} - 1} ds = \frac{1}{n} s^{\frac{1-n}{n}} ds \quad . \quad . \quad . \quad . \quad . \quad (3)$$

Replacing y by its value, and remembering that

$$s^{1-n} = \frac{1}{s^{n-1}}, \text{ we have,}$$

$$d\sqrt[n]{s} = \frac{1}{n} \frac{ds}{\sqrt[n]{s^{n-1}}} \quad . \quad . \quad . \quad . \quad . \quad (4)$$

That is, the differential of a radical of the nth degree is equal to *the differential of the quantity under the radical sign, divided by n times the* $(n - 1)^{\text{th}}$ *power of the radical.*

Cor.—If $n = 2$, we have,

$$d\sqrt{s} = \frac{ds}{2\sqrt{s}} \quad . \quad . \quad . \quad . \quad . \quad (5)$$

That is, the differential of the square root of a quantity is equal to *the differential of the quantity under the radical sign, divided by twice the radical.*

The preceding rules are sufficient to differentiate any algebraic function whatever.

EXAMPLES.

1. Let $y = 5x^3$.

This is the product of the function x^3 by the constant 5; hence, (Art. 9),

$$dy = 5d(x^3) = 5 \times 3x^2dx = 15x^2dx \; . \; . \; . \; . \; . \; \text{(Art. 12).}$$

2. Let $y = x^3 + 2x^2 + 3x + 4$.

This is a polynomial. Hence, from Art. 9,

$$dy = 3x^2dx + 4xdx + 3dx.$$

3. Let $y = (a + bx)^3$.

Considering $a + bx$ as a single quantity, we have, (Art. 12),

$$dy = 3(a+bx)^2\,d(a + bx) = 3(a + bx)^2bdx = 3b(a + bx)^2dx$$

4. Let $y = (a + bx^2)^n$.

$$dy = n(a + bx^2)^{n-1}d(a + bx^2) = 2bnx(a + bx^2)^{n-1}dx.$$

5. Let $y = \sqrt{a^2 - x^2} = (a^2 - x^2)^{\frac{1}{2}}$.

$$dy = \tfrac{1}{2}(a^2 - x^2)^{-\frac{1}{2}} \times -2xdx = -\frac{xdx}{\sqrt{a^2 - x^2}}.$$

6. Let $y = (a + x)\,(b + 2x^2)$.

By Art. 10, we have,

$$dy = (a + x)d(b + 2x^2) + (b + 2x^2)d(a + x)$$

$$= (a + x)4xdx + (b + 2x^2)dx \text{ ; or, } dy$$

$$= (6x^2 + 4ax + b)dx.$$

7. Let $y = \dfrac{ax}{a^2 + x^2}$.

By Art. 11, we have,

$$dy = \frac{(a^2 + x^2)d(ax) - axd(a^2 + x^2)}{(a^2 + x^2)^2} = \frac{a(a^2 - x^2)}{(a^2 + x^2)^2}dx.$$

8. Let $y = \sqrt{x + \sqrt{a^2 + x^2}}$.

$$dy = d\left[x + (a^2 + x^2)^{\frac{1}{2}}\right]^{\frac{1}{2}} = \frac{1}{2}\left[x + (a^2 + x^2)^{\frac{1}{2}}\right]^{-\frac{1}{2}}$$

$$\times\, d\left[x + (a^2 + x^2)^{\frac{1}{2}}\right]$$

but, $d\left[x + (a^2 + x^2)^{\frac{1}{2}}\right] = dx + xdx(a^2 + x^2)^{-\frac{1}{2}}$

$$= \left[x + (a^2 + x^2)^{\frac{1}{2}}\right](a^2 + x^2)^{-\frac{1}{2}}\,dx.$$

$$\therefore\ dy = \frac{\sqrt{x + \sqrt{a^2 + x^2}}}{2\sqrt{a^2 + x^2}}dx.$$

9. $y = ax^{\frac{5}{2}} - bx^{\frac{3}{2}} \pm c$. *Ans.* $dy = (\frac{5}{2}ax^{\frac{3}{2}} - \frac{3}{2}bx^{\frac{1}{2}})dx$.

10. $y = ax^{-\frac{5}{2}} - bx^{-\frac{3}{2}} \pm c$.

Ans. $dy = -\left(\frac{5}{2}ax^{-\frac{7}{2}} - \frac{3}{2}bx^{-\frac{5}{2}}\right)dx$.

11. $y = (a^2 - x^2)^2$. *Ans.* $dy = -4x(a^2 - x^2)dx$.

12. $y = (2ax - x^2)^3$. *Ans.* $dy = 6(a - x)(2ax - x^2)^2\,dx$.

13. $y = (a + bx^n)^m$. *Ans.* $dy = mnbx^{n-1}(a + bx^n)^{m-1}dx$.

14. $y = (2ax + x^2)^n$. *Ans.* $dy = 2n(a + x)(2ax + x^2)^{n-1}dx$.

15. $y = x(a + x)(a^2 + x^2)$.

Ans. $dy = (a^3 + 2a^2x + 3ax^2 + 4x^3)dx$.

16. $y = (a + x)^m (b + x)^n$.

Ans. $dy = (a + x)^m (b + x)^n\left[\frac{m}{a + x} + \frac{n}{b + x}\right]dx$.

17. $y = \frac{a - x}{a + x}$. *Ans.* $dy = -\frac{2a}{(a + x)^2}dx$.

18. $y = \frac{(x + 4)^2}{x + 3}$. *Ans.* $dy = \frac{(x + 2)(x + 4)}{(x + 3)^2}dx$.

19. $y = \frac{x^2 - x + 1}{x^2 + x - 1}$. *Ans.* $dy = \frac{2x(x - 2)}{(x^2 + x - 1)^2}dx$

20. $y = \sqrt{a + x}$. *Ans.* $dy = \frac{dx}{2\sqrt{a + x}}$.

21. $y = \sqrt{1 + x^2}$ *Ans.* $dy = \frac{xdx}{\sqrt{1 + x^2}}$.

22. $y = \sqrt{ax^2 + bx + c}$. *Ans.* $dy = \frac{(2ax + b)dx}{2\sqrt{ax^2 + bx + c}}$.

23. $y = (a - x)\sqrt{a + x}$. *Ans.* $dy = -\frac{(a + 3x)dx}{2\sqrt{a + x}}$.

24. $y = (a + x)\sqrt{a - x}$ *Ans.* $dy = \frac{(a - 3x)dx}{2\sqrt{a - x}}$.

25. $y = \left(1 - x^{\frac{1}{2}} + x^{\frac{2}{3}}\right)^{\frac{3}{4}}$. *Ans.* $dy = \dfrac{1}{8} \dfrac{(4x^{\frac{1}{6}} - 3)dx}{x^{\frac{1}{2}}(1 - x^{\frac{1}{2}} + x^{\frac{2}{3}})^{\frac{1}{4}}}$.

26. $y = (a^2 - x^2)\sqrt{a + x}$.

Ans. $dy = [\frac{1}{2}(a - 5x)\sqrt{a + x}]dx$.

27. $y = \dfrac{a + x}{\sqrt{a - x}}$. *Ans.* $dy = \dfrac{(3a - x)dx}{2(a - x)^{\frac{3}{2}}}$.

28. $y = \dfrac{\sqrt{a + x}}{\sqrt{a - x}}$. *Ans.* $dy = \dfrac{a dx}{(a - x)\sqrt{a^2 - x^2}}$.

29. $y = \sqrt{x - \sqrt{a^2 - x^2}}$.

Ans. $dy = \dfrac{(x + \sqrt{a^2 - x^2})dx}{2\sqrt{a^2 - x^2}(x - \sqrt{a^2 - x^2})^{\frac{1}{2}}}$.

30. $y = x(1 + x^2)(1 - x^2)^{\frac{1}{2}}$. *Ans.* $dy = \dfrac{(1 + x^2 - 4x^4)dx}{\sqrt{1 - x^2}}$.

31. $y = (1 + 2x^2)(1 + 4x^3)$.

Ans. $dy = 4x(1 + 3x + 10x^3)dx$.

32. $y = \dfrac{x}{x + \sqrt{1 - x^2}}$. *Ans.* $dy = \dfrac{dx}{2x(1 - x^2) + \sqrt{1 - x^2}}$.

33. $y = (1 + x)\sqrt{1 - x}$. *Ans.* $dy = \dfrac{(1 - 3x)dx}{2\sqrt{1 - x}}$.

34. $y = \dfrac{x^3}{\sqrt{(1 - x^2)^3}}$. *Ans.* $dy = \dfrac{3x^2 dx}{(1 - x^2)^{\frac{5}{2}}}$.

35. $y = \sqrt{\dfrac{1-\sqrt{x}}{1+\sqrt{x}}}$. *Ans.* $dy = -\dfrac{dx}{2(1+\sqrt{x})\sqrt{x-x^2}}$.

36. $y = \sqrt{2x-1-\sqrt{2x-1-\sqrt{2x-1-\text{etc., } ad\ inf.}}}$

Squaring, $y^2 = 2x-1-\sqrt{2x-1-\sqrt{2x-1-\text{etc., } ad\ inf.}}$

Hence, $y^2 = 2x-1-y$, or $y^2+y = 2x-1$.

Solving, $y = -\dfrac{1}{2} \pm \sqrt{2x-1+\tfrac{1}{4}} = -\dfrac{1}{2} \pm \dfrac{1}{2}\sqrt{8x-3}$.

$$\therefore\ dy = \pm\frac{1}{2}\frac{4dx}{\sqrt{8x-3}} = \pm\frac{2dx}{\sqrt{8x-3}}.\quad Ans.$$

III. Differentiation of Transcendental Functions.

Definition of a Transcendental Function.

14. A *transcendental function* is one in which the relation between the function and its variable cannot be expressed by the *ordinary* operations of algebra.

Transcendental functions are divided into *three* classes: *Logarithmic*, in which the relation is expressed by logarithmic symbols; *Exponential*, in which the variable enters an exponent; and *Circular*, in which the relation is expressed by means of trigonometric symbols. Thus,

$y = \log\,(ax^2 + b)$, is a *logarithmic* function,

$y = (ax^2 + c)^{2x-c}$, is an *exponential* function,

and $y = \sin^2 x - 2\tan x$, is a *trigonometric* function.

Differential of a Logarithm.

15. It is shown in Algebra that

$$\log(1 + y) = M\left(y - \frac{y^2}{2} + \frac{y^3}{3} - \frac{y^4}{4} + \text{etc.}\right) \quad \ldots . \quad (1)$$

In which M is the modulus of the system and y any quantity whatever.

Substitute for y the quantity $\frac{ds}{s}$, s being a function of x, and we have, after reduction,

$$\log\left(\frac{s + ds}{s}\right) = M\left(\frac{ds}{s} - \frac{ds^2}{2s^2} + \frac{ds^3}{3s^3} - \text{etc.}\right) \quad \ldots . . \quad (2)$$

But the logarithm of a quotient, is equal to the logarithm of the dividend, diminished by the logarithm of the divisor; changing the form of the first member and suppressing all the terms in the second member, after the first, (Art. 6), we have,

$$\log(s + ds) - \log s = M\frac{ds}{s} \quad \ldots . . \quad (3)$$

But the first member is the difference between two consecutive values of $\log s$; it is therefore the differential of the logarithm of s; hence,

$$d(\log s) = M\frac{ds}{s} \quad \ldots . . \quad (4)$$

That is, the differential of the logarithm of a quantity is equal to *the modulus into the differential of the quantity, divided by the quantity.*

In analysis, Napierian logarithms are almost always used.

Denoting the Napierian logarithm by l, and remembering that the modulus of this system is 1, we have,

$$d(ls) = \frac{ds}{s} \quad . \; . \; . \; . \; . \quad (5)$$

The base of the Napierian system is represented by the letter e, and its logarithm in that system is equal to 1.

Differential of an Exponential Function.

16. Let a be a constant quantity, s any function of x, and assume

$$y = a^s \quad . \; . \; . \; . \; . \quad (1)$$

Taking the logarithms of both members of (1), we have,

$$ly = sla \quad . \; . \; . \; . \; . \quad (2)$$

Differentiating (2), we have,

$$\frac{dy}{y} = la \times ds \quad . \; . \; . \; . \; . \quad (3)$$

Replacing y by its value, and reducing, we have,

$$d(a^s) = a^s lads \quad . \; . \; . \; . \; . \quad (4)$$

Hence, to differentiate a quantity formed by raising a constant quantity to a power denoted by a variable exponent, *multiply the quantity by the logarithm of the root and the differential of the exponent.*

Again, let us have,

$$y = t^s \quad . \; . \; . \; . \; . \quad (1)$$

in which both t and s are functions of x. Taking the logarithm of both members, we have,

$$ly = slt \quad . \; . \; . \; . \; . \quad (2)$$

Whence, by differentiating both members,

$$\frac{dy}{y} = lt\,ds + s\frac{dt}{t} \quad . \; . \; . \; . \; . \; (3)$$

Replacing y by its value, clearing of fractions and reducing, we have,

$$d(t^s) = t^s lt\,ds + st^{s-1}dt \quad . \; . \; . \; . \; . \; (4)$$

That is, to differentiate a quantity formed by raising a variable quantity to a power whose exponent is variable, *differentiate first as though the root alone were variable, then as though the exponent alone were variable, and take the sum of the results.*

EXAMPLES.

1. Let $y = l\left(\frac{a+x}{a-x}\right) = l\,(a+x) - l\,(a-x)$;

$$\therefore\ dy = \frac{d(a+x)}{a+x} - \frac{d(a-x)}{a-x} = \frac{dx}{a+x} + \frac{dx}{a-x}$$

$$= \frac{2a\,dx}{a^2 - x^2}. \quad \textit{Ans.}$$

2. Let $y = l\sqrt{\frac{1+x}{1-x}} = \frac{1}{2}l(1+x) - \frac{1}{2}l(1-x)$;

$$\therefore\ dy = \frac{1}{2}\frac{dx}{1+x} + \frac{1}{2}\frac{dx}{1-x} = \frac{dx}{1-x^2}. \quad \textit{Ans.}$$

3. Let $y = (a)^{e^x}$; $\therefore\ ly = e^x la$, and $\frac{dy}{y} = la\,e^x le\,dx$.

But, $le = 1$, $\therefore\ dy = (a)^{e^x} e^x la\,dx$. *Ans.*

4. Let $y = l(lx) = l^2x$.

By the rule, $dy = \frac{d(lx)}{lx} = \frac{dx}{xlx}$. *Ans.*

5. $y = x\,l(a + x)$. *Ans.* $dy = l(a + x)dx + \frac{xdx}{a + x}$.

6. $y = l(a + x)^2 = 2l(a + x)$. *Ans.* $dy = \frac{2dx}{a + x}$.

7. $y = e^x(x - 1)$. *Ans.* $dy = e^x x dx$.

8. $y = e^x(x^2 - 2x + 2)$. *Ans.* $dy = e^x x^2 dx$.

9. $y = \frac{e^x - 1}{e^x + 1}$. *Ans.* $dy = \frac{2e^x}{(e^x + 1)^2}dx$.

10. $y = e^x lx$. *Ans.* $dy = e^x\left(lx + \frac{1}{x}\right)dx$.

11. $y = l(x + a + \sqrt{2ax + x^2})$. *Ans.* $dy = \frac{dx}{\sqrt{2ax + x^2}}$.

12. $y = \frac{e^x}{1 + x}$. *Ans.* $dy = \frac{xe^x dx}{(1 + x)^2}$.

13. $y = l\frac{x}{\sqrt{x^2 + 1} + x}$. *Ans.* $dy = \frac{dx}{x} - \frac{dx}{\sqrt{x^2 + 1}}$.

14. $y = e^x\sqrt{\frac{1 + x}{1 - x}}$. *Ans.* $dy = e^x\frac{2 - x^2}{1 - x^2}\sqrt{\frac{1 + x}{1 - x}}dx$.

15. $y = x^x$. *Ans.* $dy = x^x(lx + 1)dx$.

16. $y = a^{lx}$. *Ans.* $dy = a^{lx}\,la\,x^{-1}dx$.

Differentials of Circular Functions.

17. Assume the equation,

$$y = \sin x \; . \; . \; . \; . \; . \; (1)$$

Giving to x the increment dx, and developing, we have,

$$y + dy = \sin(x + dx)$$

$$= \sin x \cos dx + \cos x \sin dx \; . \; . \; . \; . \; (2)$$

Because the arc dx is infinitely small, its *cosine* is equal to 1, and its *sine* is equal to the arc itself. Hence,

$$y + dy = \sin x + \cos x dx \; . \; . \; . \; . \; . \; (3)$$

Subtracting (1) from (3), and replacing y by its value, we have,

$$d(\sin x) = \cos x dx \; . \; . \; . \; . \; . \; (a)$$

Equation (a) is true for all values of x; we may therefore replace x by $(90° - x)$; this gives,

$$d[\sin(90° - x)] = \cos(90° - x) d(90° - x) \; . \; . \; . \; . \; . \; (4)$$

Reducing, we have,

$$d(\cos x) = - \sin x dx \; . \; . \; . \; . \; . \; (b)$$

We have, from trigonometry, the relation,

$$\tan x = \frac{\sin x}{\cos x} \; . \; . \; . \; . \; . \; (5)$$

Differentiating both members, we have,

$$d(\tan x) = \frac{\cos x d(\sin x) - \sin x d(\cos x)}{\cos^2 x}.$$

Performing indicated operations, and reducing by the relation, $\sin^2 x + \cos^2 x = 1$, we have,

$$d(\tan x) = \frac{dx}{\cos^2 x} \quad \ldots\ldots \quad (c)$$

Replacing x, in formula (c), by $(90° - x)$ and reducing, we have,

$$d(\cot x) = -\frac{dx}{\sin^2 x} \quad \ldots\ldots \quad (d)$$

We have, from trigonometry,

$$\text{ver-sin}x = 1 - \cos x \quad \ldots\ldots \quad (6)$$

$$\text{co-versin}x = 1 - \sin x \quad \ldots\ldots \quad (7)$$

Differentiating (6) and (7), we have,

$$d(\text{ver-sin}x) = \sin x dx \quad \ldots\ldots \quad (e)$$

$$d(\text{co-versin}x) = -\cos x dx \quad \ldots\ldots \quad (f)$$

We have, from trigonometry,

$$\sec x = \frac{1}{\cos x} \quad \ldots\ldots \quad (8)$$

$$\text{cosec}x = \frac{1}{\sin x} \quad \ldots\ldots \quad (9)$$

Differentiating (8) and (9), we have, after reduction,

$$d(\sec x) = \tan x \sec x dx \quad \ldots\ldots \quad (g)$$

$$d(\text{cosec}x) = -\cot x\, \text{cosec}x dx \quad \ldots\ldots \quad (h)$$

The *lettered* formulas, from (a) to (h) inclusive, are sufficient for the differentiation of any direct circular function.

EXAMPLES.

1. Let $y = \sin mx$.

$$dy = \cos mx \, d(mx) = m \cos mx \, . \, dx.$$

2. Let $y = \sin^m x$.

$$dy = m \sin^{m-1} x \, d\sin x = m \sin^{m-1} x \cos x \, dx.$$

3. Let $y = \sin 2x \cos x$.

$$dy = (d \sin 2x)(\cos x) + (d\cos x)(\sin 2x)$$
$$= 2\cos 2x \cos x \, dx - \sin 2x \sin x \, dx.$$

4. Let $y = l(\sin^2 x) = 2l \sin x$.

$$dy = 2 \frac{d\sin x}{\sin x} = \frac{2\cos x}{\sin x} dx = 2\cot x \, dx.$$

5. Let $y = l\left(\frac{1 + \cos x}{1 - \cos x}\right)^{\frac{1}{2}} = \frac{1}{2}l(1 + \cos x) - \frac{1}{2}l(1 - \cos x)$.

$$dy = -\frac{1}{2}\frac{\sin x dx}{1+\cos x} - \frac{1}{2}\frac{\sin x dx}{1 - \cos x} = -\frac{\sin x dx}{1 - \cos^2 x} = -\frac{dx}{\sin x}.$$

6. Let $y = e^x \cos x$.

$$dy = e^x d(\cos x) + \cos x \, d(e^x) = e^x(\cos x - \sin x) \, dx.$$

7. $y = x\sin x + \cos x$. *Ans.* $dy = x\cos x \, dx$.

8. $y = 2x\sin x + (2 - x^2)\cos x$. *Ans.* $dy = x^2 \sin x \, dx$.

9. $y = \tan x - x$. *Ans.* $dy = \tan^2 x \, dx$.

10. $y = e^{\cos x}\sin x$. *Ans.* $dy = e^{\cos x}(\cos x - \sin^2 x) dx$.

11. $y = 2l(\sin x) + \operatorname{cosec} x$. *Ans.* $dy = \left(2\cot x - \frac{\cot x}{\sin x}\right) dx$

12. $y = l(\cos x + \sqrt{-1}\sin x)$. *Ans.* $dy = \sqrt{-1}dx$

13. $y = l\left(\frac{1 + \sin x}{1 - \sin x}\right)^{\frac{1}{2}}$ *Ans.* $dy = \frac{dx}{\cos x}$.

14. $y = l[\tan(45° + \frac{1}{2}x)]$. *Ans.* $dy = \frac{dx}{\cos x}$.

15. $y = \sin(lx)$. *Ans.* $dy = \frac{1}{x}\cos(lx)dx$.

Differentials of Inverse Circular Functions.

18. It is often convenient to regard an *arc* as a function of one of its *trigonometrical lines.* Such functions are called *inverse circular functions,* and are expressed by such symbols as the following:

$$\sin^{-1}y,\ \cos^{-1}y,\ \tan^{-1}y, \text{etc.},$$

which are read *the arc whose sine is y, the arc whose cosine is y, the arc whose tangent is y,* etc.

Formulas for the differentiation of inverse circular functions may be deduced from the lettered formulas of the last article. Thus, from formula (a), we find,

$$dx = \frac{d(\sin x)}{\cos x} \quad \ldots\ldots (1)$$

If we make $\sin x = y$, we have,

$$x = \sin^{-1}y, \text{ and } \cos x = \sqrt{1 - y^2}.$$

Substituting these in (1), we have,

$$d(\sin^{-1}y) = \frac{dy}{\sqrt{1 - y^2}} \quad \ldots\ldots (a^1)$$

In like manner, from formula (b), we find,

$$dx = -\frac{d(\cos x)}{\sin x} \quad \ldots\ldots \quad (2)$$

Making $\cos x = y$, we have,

$$x = \cos^{-1} y, \text{ and } \sin x = \sqrt{1 - y^2};$$

Substituting these in (2), we have,

$$d(\cos^{-1} y) = -\frac{dy}{\sqrt{1 - y^2}} \quad \ldots\ldots \quad (b^1)$$

By a similar course of reasoning we find from the remaining lettered formulas of Art. 17, the corresponding formulas for inverse circular functions, as given below:

$$d(\tan^{-1} y) = \frac{dy}{1 + y^2} \quad \ldots\ldots \quad (c^1)$$

$$d(\cot^{-1} y) = -\frac{dy}{1 + y^2} \quad \ldots\ldots \quad (d^1)$$

$$d(\text{versin}^{-1} y) = \frac{dy}{\sqrt{2y - y^2}} \quad \ldots\ldots \quad (e^1)$$

$$d(\text{coversin}^{-1} y) = -\frac{dy}{\sqrt{2y - y^2}} \quad \ldots\ldots \quad (f^1)$$

$$d(\sec^{-1} y) = \frac{dy}{y\sqrt{y^2 - 1}} \quad \ldots\ldots \quad (g^1)$$

$$d(\text{cosec}^{-1} y) = -\frac{dy}{y\sqrt{y^2 - 1}} \quad \ldots \quad (h^1)$$

EXAMPLES.

1. Let $y = \cos^{-1} x\sqrt{1-x^2} = \cos^{-1}\sqrt{x^2 - x^4}$.

$$d\sqrt{x^2 - x^4} = \frac{(1-2x^2)dx}{\sqrt{1-x^2}}$$

$$\text{and } \sqrt{1-(x^2 - x^4)} = \sqrt{1 - x^2 + x^4}.$$

$$\therefore dy = -\frac{(1-2x^2)\,dx}{\sqrt{(1-x^2)(1-x^2+x^4)}}.$$

2. Let $y = \sin^{-1}\dfrac{x}{\sqrt{1+x^2}}$.

$$dy = d\left(\frac{x}{\sqrt{1+x^2}}\right) \div \sqrt{1 - \frac{x^2}{1+x^2}}$$

$$= \frac{dx}{(1+x^2)^{\frac{3}{2}}} \div \frac{1}{(1+x^2)^{\frac{1}{2}}} = \frac{dx}{(1+x^2)}.$$

3. Let $y = \sin^{-1} 2x\sqrt{1-x^2}$.

$$dy = d(2x\sqrt{1-x^2}) \div \sqrt{1 - 4x^2(1-x^2)}$$

$$= \frac{2(1-2x^2)dx}{\sqrt{1-x^2}} \div (1-2x^2) = \frac{2dx}{\sqrt{1-x^2}}.$$

4. $y = \tan^{-1}\dfrac{2x}{1-x^2}$. *Ans.* $dy = \dfrac{2dx}{1+x^2}$

5. $y = \sin^{-1}\dfrac{1-x^2}{1+x^2}$. *Ans.* $dy = -\dfrac{2dx}{1+x^2}$

6. $y = \sqrt{a^2 - x^2} + a\sin^{-1}\frac{x}{a}$. *Ans.* $dy = \left(\frac{a - x}{a + x}\right)^{\frac{1}{2}} dx$.

7. $y = \frac{1}{2}\operatorname{versin}^{-1}\frac{8x}{9}$. *Ans.* $dy = \frac{dx}{\sqrt{9x - 4x^2}}$

8. $y = \sec^{-1} 2x$. *Ans.* $dy = \frac{dx}{x\sqrt{4x^2 - 1}}$

IV. Successive Differentiation and Development of Functions.

Successive Differentials.

19. The differential obtained immediately from the function is *the first differential of the function;* the differential of the first differential is *the second differential of the function;* the differential of the second differential is *the third differential of the function,* and so on; differentials thus obtained are called *successive differentials,* and the operation of obtaining them is called *successive differentiation.* If a function be denoted by y, its successive differentials will be denoted by the symbols dy, d^2y, d^3y, etc. Thus, if $y = ax^3$, we have, by successive differentiation, dx being constant, $dy = 3ax^2dx$, $d^2y = 6axdx^2$, $d^3y = 6adx^3$. $d^4y = 0$.

Successive Differential Coefficients

20. If the differential of a function be divided by the differential of the variable, the quotient is *the first differential coefficient of the function;* the differential coefficient of the first differential coefficient is the *second differential coefficient* of the function, and so on. Differential coefficients, derived in this manner, are called *successive differential coefficients.* If a function of x be denoted by y, its successive differential coefficients are denoted by the symbols, $\frac{dy}{dx}$, $\frac{d^2y}{dx^2}$, $\frac{d^3y}{dx^3}$, etc. Thus, if we have, as before, $y = ax^3$, we have, from the principles just explained,

$$\frac{dy}{dx} = 3ax^2\,; \ \frac{d^2y}{dx^2} = 6ax\,; \ \frac{d^3y}{dx^3} = 6a\,; \ \frac{d^4y}{dx^4} = 0.$$

It is to be observed that the successive differential coefficients are entirely independent of the differential of the variable; so long therefore as this is infinitesimal, the differential coefficients will in no way be affected by any change in its absolute value.

EXAMPLES.

Find the successive differential coefficients of the following functions:

1. $y = ax^n$.

Ans. $\frac{dy}{dx} = nax^{n-1}\,; \ \frac{d^2y}{dx^2} = n(n-1)ax^{n-2}\,;$

$$\frac{d^3y}{dx^3} = n(n-1)(n-2)ax^{n-3}\,; \text{ etc}$$

If n is a positive whole number, there will be a finite number of successive differential coefficients; otherwise their number is infinite.

2. $y = ax^3 + bx^2$.

Ans. $\frac{dy}{dx} = 3ax^2 + 2bx\,; \ \frac{d^2y}{dx^2} = 6ax + 2b\,;$

$$\frac{d^3y}{dx^3} = 6a\,; \ \frac{d^4y}{dx^4} = 0.$$

3. $y = a^x$.

Ans. $\frac{dy}{dx} = a^x \, la$; $\frac{d^2y}{dx^2} = a^x(la)^2$; $\frac{d^3y}{dx^3} = a^x(la)^3$

$$\ldots\ldots \frac{d^nx}{dx^n} = a^x(la)^n.$$

4. $y = \sin x$.

Ans. $\frac{dy}{dx} = \cos x$; $\frac{d^2y}{dx^2} = -\sin x$;

$$\frac{d^3y}{dx^3} = -\cos x; \quad \frac{d^4y}{dx^4} = \sin x; \text{ etc.}$$

5. $y = l(x + 1)$.

Ans. $\frac{dy}{dx} = \frac{1}{x+1} = (x + 1)^{-1}$; $\frac{d^2y}{dx^2} = -(x + 1)^{-2}$;

$$\frac{d^3y}{dx^3} = 2(x + 1)^{-3}; \quad \frac{d^4y}{dx^4} = -6(x + 1)^{-4}; \text{ etc.}$$

6. $y = xe^x$.

Ans. $\frac{dy}{dx} = (x + 1)e^x$; $\frac{d^2y}{dx^2} = (x + 2)e^x$

$$\ldots\ldots \frac{d^ny}{dx^n} = (x + n)e^x.$$

McLaurin's Formula.

21. McLaurin's formula, is a formula for developing a function of one variable into a series arranged according to the ascending powers of that variable, the coefficients being constant. Let y be any function of x, and assume the development,

$$y = A + Bx + Cx^2 + Dx^3 + Ex^4 + \text{etc.} \ldots\ldots (1)$$

It is required to find such values for A, B, C, etc., as will make the assumed development true for all values of x. Differentiating, and finding the successive differential co efficients of y, we have,

$$\frac{dy}{dx} = B + 2Cx + 3Dx^2 + 4Ex^3 + \text{etc.} \quad \ldots \; (2)$$

$$\frac{d^2y}{dx^2} = 1.2C + 1.2.3Dx + 3.4Ex^2 + \text{etc.} \quad \ldots \; (3)$$

$$\frac{d^3y}{dx^3} = 1.2.3D + 1.2.3.4Ex + \text{etc.} \quad \ldots\ldots \; (4)$$

etc., etc., etc.

But, by hypothesis, the value of y, and consequently the values of its successive differential coefficients, are to be true for all values of x; hence, they must be so for $x = 0$. Making $x = 0$, in equations (1), (2), (3), etc., and denoting what y becomes under that hypothesis by (y); what $\frac{dy}{dx}$ becomes by $\left(\frac{dy}{dx}\right)$; what $\frac{d^2y}{dx^2}$ becomes by $\left(\frac{d^2y}{dx^2}\right)$, and so on; we have,

$$(y) = A; \qquad \therefore\ A = (y);$$

$$\left(\frac{dy}{dx}\right) = B; \qquad \therefore\ B = \left(\frac{dy}{dx}\right);$$

$$\left(\frac{d^2y}{dx^2}\right) = 1.2C; \qquad \therefore\ C = \frac{1}{1.2}\left(\frac{d^2y}{dx^2}\right);$$

$$\left(\frac{d^3y}{dx^3}\right) = 1.2.3.D; \quad \therefore\ D = \frac{1}{1.2.3}\left(\frac{d^3y}{dx^3}\right);$$

etc., etc., etc.

Substituting these values in (1), we have,

$$y = (y) + \left(\frac{dy}{dx}\right) \cdot \frac{x}{1} + \left(\frac{d^2y}{dx^2}\right)\frac{x^2}{1.2} + \left(\frac{d^3y}{dx^3}\right) \cdot \frac{x^3}{1.2.3} + \text{etc.} \,.\, (5)$$

which is the formula required. Hence, to develop a function of one variable in terms of that variable, find its successive differential coefficients; then make the variable equal to 0 in the function and its successive differential coefficients, and substitute the results for (y), $\left(\frac{dy}{dx}\right)$, $\left(\frac{d^2y}{dx^2}\right)$, etc., in formula (5).

Thus, let it be required to develop $\sin x$ into a series arranged with reference to x. We have,

$$y = \sin x, \quad \therefore \ (y) = 0$$

$$\frac{dy}{dx} = \cos x, \quad \therefore \ \left(\frac{dy}{dx}\right) = 1$$

$$\frac{d^2y}{dx^2} = -\sin x \quad \therefore \ \left(\frac{d^2y}{dx^2}\right) = 0$$

$$\frac{d^3y}{dx^3} = -\cos x, \quad \therefore \ \left(\frac{d^3y}{dx^3}\right) = -1$$

etc., etc., etc.

Hence,

$$\sin x = x - \frac{x^3}{1.2.3} + \frac{x^5}{1.2.3.4.5} - \frac{x^7}{1.2.3.4.5.6.7} + \text{etc.} \ldots (6)$$

In like manner we find,

$$\cos x = 1 - \frac{x^2}{1.2} + \frac{x^4}{1.2.3.4} - \frac{x^6}{1.2.3.4.5.6} + \text{etc.} \ldots\ldots (7)$$

It is to be observed that (7) may be found from (6) by differentiating and then dividing by dx.

McLaurin's Formula enables us to develop any function of one variable when it can be developed in accordance with the assumed law. But there are functions which cannot be developed according to the ascending powers of the variable; in this case the function, or some of its successive differential coefficients, become ∞, when the variable is made equal to 0.

As a general rule, when the application of a formula gives an infinite result, the formula is inapplicable in that particular case.

EXAMPLES.

1. $y = (a + x)^n$.

$$Ans.\ y = a^n + \frac{n}{1}a^{n-1}x + \frac{n(n-1)}{1.2}a^{n-2}x^2$$

$$+ \frac{n(n-1)(n-2)}{1.2.3}a^{n-3}x^3 + \text{etc.}$$

2. $y = l(1 + x)$.

$$Ans.\ y = x - \frac{x^2}{2} + \frac{x^3}{3} - \frac{x^4}{4} + \frac{x^5}{5} - \text{etc.}$$

3. $y = a^x$.

$$Ans.\ y = 1 + \frac{(la)}{1}x + \frac{(la)^2}{1.2}x^2 + \frac{(la)^3}{1.2.3}x^3 + \frac{(la)^4}{1.2.3.4}x^4 + \text{etc.}$$

4. $y = \sqrt{1 + x^2} = (1 + x^2)^{\frac{1}{2}}$.

Make $x^2 = z$, and devel p; then replace z by its value.

$$Ans.\ y = 1 + \frac{x^2}{2} - \frac{x^4}{8} + \frac{x^6}{16} - \frac{5x^8}{128} + \text{etc.}$$

5. $y = e^{\sin x}$

$$Ans.\ y = 1 + x + \frac{x^2}{2} - \frac{3x^4}{1.2.3.4} - \frac{8x^5}{1.2.3.4.5} - \frac{3x^6}{1.2.3.4.5.6} + \text{etc.}$$

6. $y = xe^x + e^x - 1$.

$$Ans.\ y = 2x + \frac{3x^2}{1.2} + \frac{4x^3}{1.2.3} + \frac{5x^4}{1.2.3.4} + \text{etc.}$$

Taylor's Formula.

22. Taylor's Formula is a formula for developing a function of the *sum* of two variables into a series arranged according to the ascending powers of one, with coefficients that are functions of the other.

LEMMA.

If $u' = f(x + y)$, the differential coefficient of u' will be the same, whether we suppose x to remain constant and y to vary, or y to remain constant and x to vary, for the form of the function is the same, whichever we suppose to vary; and it has been shown, in Art. 20, that the value of the differential coefficient is independent of the value of the differential of the variable. Hence, if $x + y$ be increased either by dx or by dy, and the differential coefficient taken, the result will be the same, which was to be shown.

Let u' be a function of $(x + y)$, and assume the development,

$$u' = P + Qx + Rx^2 + Sx^3 + Tx^4 + \text{etc.} \quad . \quad . \quad . \quad . \quad . \quad (1)$$

in which P, Q, R, etc., are functions of y. It is required to find such values of P, Q, R, etc., as will make equation

(1) true for all values of both x and y. Since the assumed development is to be true for all values of x and y, it must be so for $x = 0$. Making $x = 0$ in (1), and denoting what u' becomes under this hypothesis, by u, we find,

$$P = u.$$

That is, P is what the original function becomes, when the *leading* variable is made equal to 0.

Finding the differential coefficient of u', under the supposition that x is constant and y variable, we have,

$$\frac{du'}{dy} = \frac{dP}{dy} + \frac{dQ}{dy}x + \frac{dR}{dy}x^2 + \frac{dS}{dy}x^3 + \text{etc.} \quad . \quad . \quad . \quad . \quad . \quad (2)$$

Again, finding the differential coefficient of u', on the supposition that y is constant and x variable, we have,

$$\frac{du'}{dx} = Q + 2Rx + 3Sx^2 + 4Tx^3 + \text{etc.} \quad . \quad . \quad . \quad . \quad . \quad (3)$$

But, the first members of (2) and (3) are equal, by the lemma; hence their second members are also equal. If we place them equal we have an identical equation, because it is true for all values of x, and consequently the coefficients of the like powers of x in the two members are equal to each other. Placing their coefficients equal, we have the following results:

$$Q = \frac{dP}{dy} \qquad \therefore\ Q = \frac{du}{dy},$$

$$2R = \frac{dQ}{dy} \qquad \therefore\ R = \frac{1}{1.2}\frac{d^2u}{dy^2},$$

$$3S = \frac{dR}{dy} \qquad \therefore\ S = \frac{1}{1.2.3}\frac{d^3u}{dy^3},$$

etc., etc., etc.

Substituting the values of P, Q, R, S, etc., in (1), we have,

$$u' = u + \frac{du}{dy} \cdot \frac{x}{1} + \frac{d^2u}{dy^2} \cdot \frac{x^2}{1.2} + \frac{d^3u}{dy^3} \cdot \frac{x^3}{1.2.3}$$

$$+ \frac{d^4u}{dy^4} \cdot \frac{x^4}{1.2.3.4} + \text{etc.} \quad . \; . \; . \; . \; . \quad (4)$$

which is the required formula. Hence, to develop any function of the sum of two variables, we make the leading variable equal to 0, and find the successive differential coefficients of the result; then substitute them in the formula.

Thus, let it be required to develop $(x + y)^n$ into a series arranged according to the ascending powers of y. Making $y = 0$, we have,

$$u \quad = x^n,$$

$$\frac{du}{dx} = nx^{n-1},$$

$$\frac{d^2u}{dx^2} = n(n - 1)x^{n-2},$$

$$\frac{d^3u}{dx^3} = n(n - 1)\,(n - 2)x^{n-3},$$

etc., etc., etc.

Substituting these in (4), we have,

$$(x + y)^n = x^n + \frac{n}{1}x^{n-1}y + \frac{n.n - 1}{1.2}x^{n-2}y^2$$

$$+ \frac{n(n - 1\,(n - 2)}{1.2.3}x^{n-3}y^3 + \text{etc.}$$

This is the binomial formula, in which n is any constant.

The formula is applicable to every function of the sum of two variables; but it sometimes happens that certain values of the variable entering the coefficients make the first term or some of its successive differential coefficients *infinite;* for these particular values, the function cannot be expressed by a series of the proposed form.

EXAMPLES.

Develop the following functions in terms of y.

1. $u' = \sin(x + y)$.

Making $y = 0$, we have, $u = \sin x$;

$$\therefore \frac{du}{dx} = \cos x; \quad \frac{d^2u}{dx^2} = -\sin x; \text{ etc.};$$

hence, $u' = \sin x + \cos x \frac{y}{1} - \sin x \frac{y^2}{1.2} - \cos x \frac{y^3}{1.2.3}$

$$+ \sin x \frac{y^4}{1.2.3.4} - \text{etc.}$$

Making $x = 0$, whence $\sin x = 0$, and $\cos x = +1$, we have,

$u = \sin y = y - \frac{y^3}{1.2.3} + \frac{y^5}{1.2.3.4.5} + \text{etc.}$, as already shown.

2. $u = l(x + y)$.

$$\textit{Ans. } u' = lx + \frac{y}{x} - \frac{y^2}{2x^2} + \frac{y^3}{3x^3} - \frac{y^4}{4x^4} + \text{etc.}$$

3. $u' = a^{x+y}$.

$$\textit{Ans. } u' = a^x\left(1 + (la)y + \frac{(la)^2}{1.2}y^2 + \frac{(la)^3}{1.2.3}y^3 + \text{etc.}\right).$$

4. $u' = \cos(x + y)$.

Ans. $u' = \cos x - \sin x \frac{y}{1} - \cos x \frac{y^2}{1.2}$

$$+ \sin x \frac{y^3}{1.2.3} + \cos x \frac{y^4}{1.2.3.4} - \text{etc.}$$

V. Differentiation of Functions of two Variables, and of Implicit Functions.

Geometrical Representation of a Function of two Variables.

23. It may be shown, as in Art. 3, that every function of two variables represents the ordinate of a surface of which those variables are the corresponding abscissas. If the vertical ordinate be taken as the function, it may be expressed by the equation,

$$z = f(x, y) \quad . \; . \; . \; . \; . \quad (1)$$

In this equation, x and y are independent variables, and each may vary precisely as though the other were constant. To illustrate, let PQR be the surface whose equation is (1).

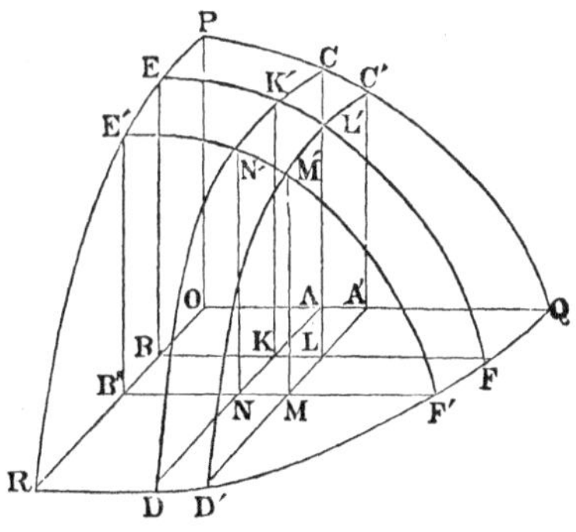

Fig. 2.

If, for a given value of y as OB, we suppose x to vary, equation (1) will represent a section of the surface parallel to the plane xz and at the distance OB from it. If for a given value of x, as OA, we suppose y to vary, equation (1) will represent a section parallel to the plane yz

and at the distance OA from it. These sections will be called *parallel sections*, and the corresponding planes will be called *planes of parallel section*. The planes ACD and BEF determine the ordinate $KK' = z$. Let ACD $A'C'D'$, be consecutive planes of parallel section, at a distance from each other equal to dx, and BEF, $B'E'F'$, also consecutive planes of parallel section, at a distance from each other equal to dy. These planes determine four ordinates KK', LL', NN', and MM', which may be denoted by the symbols z, z', z'', and z'''. Of these, z and z' are consecutive when x alone varies, z and z'' are consecutive when y alone varies, and z and z''' are consecutive when both x and y vary.

Differentials of a Function of two Variables.

24. The difference between two consecutive states of the function when x alone varies is called *the partial differential* with respect to x, and is denoted by the symbol $(dz)_x$; the difference between two consecutive states when y alone varies is called *the partial* differential with respect to y, and is denoted by the symbol $(dz)_y$; the difference between two consecutive states when both x and y vary is called the total differential, and is denoted by the ordinary symbol, dz without a subscript letter. Let us assume the figure and notation of Art. 23. We have, from the definition,

$$z' = z + (dz)_x; \text{ whence, by differentiation,}$$

$$(dz')_y = (dz)_y + (d^2z)_{x,\,y}, \quad \ldots\ldots \quad (2)$$

The symbol $(d^2z)_{x,\,y,}$ indicates the result obtained by differentiating z as though x were the only variable, and then differentiating that result as though y were the only variable, the order of the subscript letters indicating the order of differentiation. From what precedes, we also have,

$$z''' = z' + (dz')_y;$$

substituting for z' and $(dz')_y$, their values taken from (2), we have,

$$z''' = z + (dz)_x + (dz)_y + (d^2z)_{x,\,y};$$

transposing z to the first member, replacing $z''' - z$ by its value dz, and neglecting the part $(d^2z)_{x,\,y}$, because it is an infinitesimal of the second order, we have, finally,

$$dz = (dz)_x + (dz)_y \quad . \; . \; . \; . \; . \quad (3)$$

Hence, *the total differential is equal to the sum of the partial differentials.*

EXAMPLES.

1. $z = ax^2 + by^3$.

$$(dz)_x = 2axdx; \; (dz)_y = 3by^2dy \quad \therefore \; dz = 2axdx + 3by^2dy.$$

2. $z = ax^2y^3$. *Ans.* $dz = 2ay^3xdx + 3ax^2y^2dy$.

3. $z = x^y$. *Ans.* $dz = yx^{y-1}dx + x^y lxdy$.

Notation employed to designate Partial Differential Coefficients.

25. From the nature of the case, there can be no such thing as a differential coefficient of a function of two variables; but the quotient of a partial differential by the differential of the corresponding variable, is called *a partial differential coefficient.* The form of the symbol indicates the variable with reference to which the function has been differentiated, and no subscript letter is required. Thus, in Example 2, Art. 24, we have,

$$\frac{dz}{dx} = 2ay^3x, \text{ and } \frac{dz}{dy} = 3ax^2y^2.$$

A similar notation is employed for partial differential coefficients of a higher order, as will be seen in the following article.

Successive Differentiation of Functions of two Variables.

26. The second differential of a function of two variables is found by differentiating the first differential by the rules already given. The third differential comes from the second in the same manner that the second comes from the first, and so on. In finding the higher partial differentials, the result obtained by differentiating the function first with respect to x, and that result with respect to y, is the same as though we had differentiated the function first with respect to y, and the result with respect to x. For, from Art. 24, we have,

$$z''' = z' + (dz')_y = z + (dz)_x + (dz)_y + (d^2z)_{x,\,y}.$$

And in like manner, we have,

$$z''' = z'' + (dz'')_x = z + (dz)_y + (dz)_x + (d^2z)_{y,\,x}.$$

Equating these values of z''', and reducing, we find,

$$(d^2z)_{x,\,y} = (d^2z)_{y,\,x} \quad . \quad . \quad . \quad . \quad (4)$$

which was to be shown. From equation (4), by an extension of the notation in Art. 25, we have,

$$\frac{d^2z}{dxdy} = \frac{d^2z}{dydx}, \text{ or } \frac{d\left(\frac{dz}{dx}\right)}{dy} = \frac{d\left(\frac{dz}{dy}\right)}{dx} \quad \ldots\ldots \quad (5)$$

From what precedes, we have,

$$dz = (dz)_x + (dz)_y;$$

$$\therefore d^2z = (d^2z)_{x,\,x} + (d^2z)_{x,\,y} + (d^2z)_{y,\,y} + (d^2z)_{y,\,x}$$

$$= (d^2z)_{x,\,x} + 2(d^2z)_{x,\,y} + (d^2z)_{y,\,y}.$$

EXAMPLES.

1. Given, $z = x^3y^2$, to find d^2z.

$$(d^2z)_{x,\,x} = 6y^2xdx^2; \; (d^2z)_{x,\,y} = 6x^2ydxdy;$$

$$(d^2z)_{y,\,y} = 2x^3dy^2.$$

$$\therefore d^2z = 6y^2xdx^2 + 12x^2ydxdy + 2x^3dy^2.$$

2. Given, $z = y^3x^{\frac{5}{2}}$, to find d^2z.

$$\textit{Ans. } d^2z = \frac{15}{4}y^3x^{\frac{1}{2}}dx^2 + 2\left(\frac{15}{2}y^2x^{\frac{3}{2}}\right)dydx + 6x^{\frac{5}{2}}ydy^2$$

Extension to three or more Variables

27. If we have a function of three or more independent variables, we may find its differential by differentiating separately with respect to each variable, and taking the sum of the partial differentials thus obtained. The second and higher differentials are found in an entirely analogous manner.

Definition of Explicit and Implicit Functions.

28. An *explicit* function, is one in which the value of the function is directly expressed in terms of the variable. Thus, $y = \sqrt{2rx - x^2}$, is an explicit function. An *implicit* function, is one in which the value of the function is not directly given in terms of the variable. Thus, in the equation,

$$ay^2 + bxy + cx^2 + d = 0;$$

y is an implicit function of x. Implicit functions are generally connected with their variables by one or more equations. When these equations are solved the implicit function becomes explicit.

Differentials of Implicit Functions.

29. The differential of an implicit function may be found without first finding the function itself. For, if we differentiate both members of the first equation, we shall thus find a new equation, which, with the given one, will enable us to find either the differential, or the differential coefficient of the function. For example, suppose we have the equation,

$$y^2 + 2xy + x^2 - a^2 = 0 \quad . \quad . \quad . \quad . \quad (1)$$

to find the differential coefficient of y. Differentiating (1), and dividing by 2, we have,

$$ydy + xdy + ydx + xdx = 0 \ldots\ldots (2)$$

Finding the value of $\frac{dy}{dx}$, we have,

$$\frac{dy}{dx} = -1.$$

Again, let us have the relation,

$$xy = m \ldots\ldots (1)$$

to find the value of $\frac{dy}{dx}$. Differentiating (1), we have,

$$xdy + ydx = 0 \ldots\ldots (2)$$

Whence,

$$\frac{dy}{dx} = -\frac{y}{x} \ldots\ldots (3)$$

But from (1), we have,

$$\frac{y}{x} = \frac{m}{x^2}.$$

Substituting in (3), we find,

$$\frac{dy}{dx} = -\frac{m}{x^2} \ldots\ldots (4)$$

EXAMPLES.

Find the first and second differential coefficients of y in the following implicit functions:

1. $y^3 - 3y + x = 0$.

We have,

$$3y^2dy - 3dy + dx = 0, \quad \therefore \frac{dy}{dx} = \frac{1}{3} \cdot \frac{1}{1 - y^2};$$

also,

$$6ydy^2 + 3y^2d^2y - 3d^2y = 0,$$

$$\therefore \frac{d^2y}{dx^2} = \frac{2y}{1-y^2} \cdot \frac{dy^2}{dx^2} = \frac{2}{9} \cdot \frac{y}{(1-y^2)^3}.$$

2. $y^2 - 2mxy + x^2 - a = 0$.

$$\textit{Ans.}\ \frac{dy}{dx} = \frac{my - x}{y - mx}, \text{ and } \frac{d^2y}{dx^2} = \frac{a(m^2 - 1)}{(y - mx)^3}$$

PART II.

APPLICATIONS OF THE DIFFERENTIAL CALCULUS.

I. TANGENTS AND ASYMPTOTES.

Geometrical Representation of the First Differential Coefficient.

30. It has been shown (Art. 3), that any function of one variable may be represented by the ordinate of a curve, of which that variable is the corresponding abscissa. We have also seen (Art. 5), that an element of the curve, PQ, does not differ from a straight line. This line, prolonged toward T, is tangent to the curve at P. The angle that the tangent makes with the axis of X is equal to RPQ; denoting it by θ, and remembering that the tangent of the angle at the base of a right angled triangle is equal to the perpendicular divided by the base, we have,

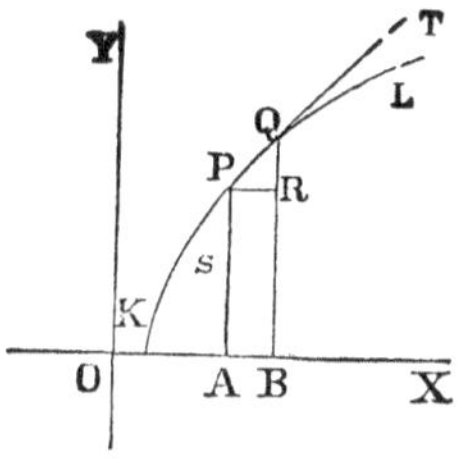

Fig. 3.

$$\tan \theta = \frac{dy}{dx}.$$

The tangent of θ is taken as the measure of the *slope*, not only of the tangent, but also of the curve at the point P. Hence, *the slope of a curve, at any point, is measured by the first differential coefficient of the ordinate at that point.*

The slope is positive or negative, that is, the curve slopes upward or downward, according as the first differential coefficient is *plus*, or *minus*.

Applications.

31. The principle just demonstrated enables us to find the point of a curve, at which the tangent makes a given angle with the axis of x.

Thus, let it be required to find the point on a given parabola at which the tangent makes an angle of 45° with the axis. Assume the equation of the parabola,

$$y^2 = 2px; \quad \therefore \quad \frac{dy}{dx} = \frac{p}{y}.$$

Placing this equal to 1, we have,

$$\frac{p}{y} = 1, \text{ or } y = p.$$

But $y = p$, is the ordinate through the focus. Hence, the required point is at the upper extremity of the ordinate through the focus.

Again, let it be required to find the point at which a tangent to the ellipse is parallel to the axis of x. Assume the equation,

$$\frac{y^2}{b^2} + \frac{x^2}{a^2} = 1; \quad \therefore \quad \frac{dy}{dx} = -\frac{b^2x}{a^2y}.$$

Placing this equal to 0, we have,

$$-\frac{b^2x}{a^2y} = 0,$$

which can only be satisfied by making $x = 0$; this, in the equation of the curve, gives $y = \pm b$. Hence, the tangent at either vertex of the conjugate axis fulfills the given condition.

Again, to find where the tangent to an hyperbola is perpendicular to the axis; assume the equation of the curve,

$$\frac{x^2}{a^2} - \frac{y^2}{b^2} = 1 \quad . \quad . \quad . \quad . \quad . \quad (1)$$

whence, by the rule,

$$\frac{dy}{dx} = \frac{b^2 x}{a^2 y} = \infty; \quad \therefore \; y = 0, \; x = \pm a.$$

Hence, the tangent at either extremity of the transverse axis, fulfills the given condition.

Equations of the Tangent and Normal.

32. Let P be a point of the curve whose co-ordinates are x'', y''; then will the equation of a straight line through it, be of the form,

$$y - y'' = a(x - x'') \quad . \quad . \quad \quad . \quad . \quad (1)$$

in which a is the slope. If we make a equal the tangent of RPQ, the line will be tangent to the curve. But the tangent of RPQ is $\frac{dy''}{dx''}$; hence, we have, for the equation of a tangent line,

$$y - y'' = \frac{dy''}{dx''}(x - x'') \quad . \quad . \quad . \quad . \quad . \quad (2)$$

If we make a equal to *minus* the reciprocal of the tangent of RPQ, the line will be perpendicular to the tangent, that is, it will be normal to the curve; hence, we have, for the equation of a normal line,

$$y - y'' = -\frac{dx''}{dy''}(x - x'') \quad . \quad . \quad . \quad . \quad . \quad (3)$$

In these equations x and y are general co-ordinates of the lines, and x'', y'' are the co-ordinates of the point of contact.

Let it be required to find the equation of a tangent to an ellipse. We found, in the last article, the value of $\frac{dy}{dx} = -\frac{b^2x}{a^2y}$; substituting for x and y the particular values x'' and y'', and putting the result in (2), we find,

$$y - y'' = -\frac{b^2x''}{a^2y''}(x - x'');$$

whence, by reduction,

$$\frac{yy''}{b^2} + \frac{xx''}{a^2} = 1 \quad . \quad . \quad . \quad . \quad . \quad (4)$$

Making $a = b$, in this equation, we have,

$$yy'' + xx'' = a^2,$$

which is the equation of a tangent to a circle.

To find the equation of a normal to an ellipse we substitute the value of $\frac{dx''}{dy''}$, in equation (3), which gives,

$$y - y'' = \frac{a^2y''}{b^2x''}(x - x'') \quad . \quad . \quad . \quad . \quad . \quad (5)$$

Making $a = b$, we have,

$$y - y'' = \frac{y''}{x''}(x - x''),$$

which is the equation of a normal to a circle.

If, in equation (2), we make $y = 0$, the corresponding value of $x - x''$ will express the length of the subtangent, counted from the foot of the ordinate through the point of contact; denoting this by $S.T.$, we have,

$$S.T = -y''\frac{dx''}{dy''} \quad . \quad . \quad . \quad . \quad . \quad (6)$$

If, in equation (3), we make $y = 0$, the corresponding value of $x - x''$ will express the length of the subnormal, counted from the foot of the ordinate through the point of contact; denoting this by $S.N.$, we have,

$$S.N = y''\frac{dy''}{dx''} \quad . \quad . \quad . \quad . \quad . \quad (7)$$

Substituting in (6) and (7) the values of $\frac{dy''}{dx''}$ taken from the equation of the ellipse, we have,

$$S.T = +\frac{a^2y''^2}{b^2x''}, \text{ and } S.N = -\frac{b^2x''}{a^2}.$$

Asymptotes.

33. An *asymptote* to a curve, is a line that continually approaches the curve and becomes tangent to it at an infinite distance. The characteristics of an asymptote are, that it is tangent to the curve at a point that is infinitely distant, and that it cuts one or both axes at a finite distance from the origin.

To ascertain whether a curve has an asymptote, assume the equation of the tangent (Art. 32), and in it make x, and y, successively equal to 0.

In this manner we find,

$$AD = y'' - x''\frac{dy''}{dx''}, \text{ and}$$

$$AT = x'' - y''\frac{dx''}{dy''} \quad \ldots\ldots (1)$$

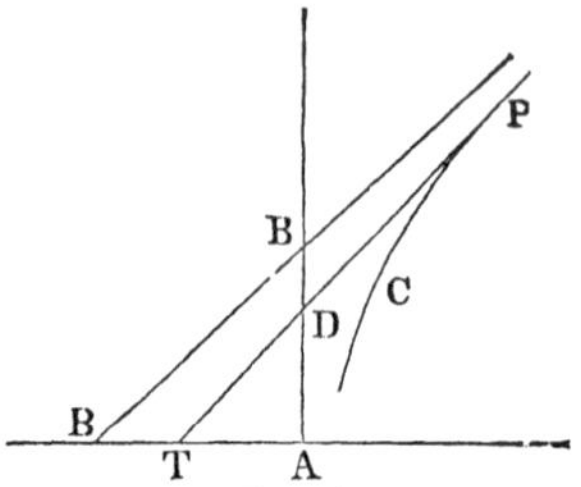

Fig. 4.

We then find $\frac{dy''}{dx''}$ from the equation of the curve, substitute it in (1) and make the hypothesis that places the point (x'', y'') at an infinite distance. If this supposition make either AD or AT finite, the curve has an asymptote, otherwise not.

Let it be required to find whether the hyperbola has an asymptote. We have found (Art. 31), $\frac{dy''}{dx''} = \frac{b^2x''}{a^2y''}$, which in (1) gives,

$$AD = y'' - \frac{b^2x''^2}{a^2y''}, \text{ and } AT = x'' - \frac{a^2y''^2}{b^2x''} \quad \ldots\ldots (2)$$

Whence, by reduction,

$$AD = \frac{a^2y''^2 - b^2x''^2}{a^2y''}, \text{ and } AT = \frac{b^2x''^2 - a^2y''^2}{b^2x''} \quad \ldots\ldots (3)$$

or, $$AD = -\frac{b^2}{y''}, \text{ and } AT = \frac{a^2}{x''} \quad \ldots\ldots (4)$$

The only hypothesis that places a point of the hyperbola at an infinite distance is $y'' = \pm\infty$, and $x'' = \pm\infty$; both these sets of values, in (4), make AD and AT equal to 0. Hence, the hyperbola has two asymptotes, both passing through the centre.

To find whether the parabola has an asymptote, make $\frac{dy''}{dx''} = \frac{p}{y''}$ in (1), whence,

$$AD = y'' - \frac{px''}{y''} = \frac{px''}{y''}, \text{ and } AT = -x''.$$

If we make $y'' = \infty$, and $x'' = \infty$, to put the point of contact at an infinite distance, we find both AD and AT equal to ∞, which shows that the parabola has no asymptote.

Order of Contact.

34. Two lines have a contact of the *first order*, when they have two consecutive points in common. The first of these is *the point of contact;* this point is common to both lines, and, as we have just seen, the first differential coefficients of the ordinates of the two lines at this point are equal. Conversely, if two lines have a common point, and if the first differential coefficients of the ordinates of the lines at that point are equal, the lines have a contact of the first order.

Two lines have a contact of the *second order*, when they have three consecutive points in common. The first of the three is *the point of contact*, and because the lines have three consecutives ordinates common, the first and second differential coefficients of their ordinates at this point are equal. Conversely, if two lines have a common point, and the first and second differential coefficients of their ordinates at that point equal, the lines have a contact of the *second order*.

In like manner it may be shown, if two lines have a point in common, and n successive differential coefficients of their ordinates at that point equal, that the lines have a contact of the nth order, n being any positive whole number.

If a line be applied so as to cut a given line in an *even number of points*, it will, just before the first, and just after the last, lie on the same side of the given line; and this is true when the points of secancy are consecutive, in which case the lines have a contact of an *odd* order. Hence, if two lines have a contact of an odd order, they do *not intersect* at the point of contact. If the applied line cut the given line in an *odd number of points*, it will, just before the first, and just after the last, lie on opposite sides of the given line; and this is also true when the points of secancy are consecutive; hence, if two lines have a contact of an even order, they *intersect* at the point of contact.

If the applied line be straight, and the contact of the nth order, the given curve will have n consecutive values of dy, equal to each other. If each of these be taken from the next in order, the differences will be 0, that is, the curve will have $n - 1$ consecutive values of d^2y equal to 0. In like manner, we may show that it has $n - 2$ consecutive values of d^3y equal to 0, and so on, to the $(n + 1)^{th}$ differential of y, which will not be 0, but will be *plus* or *minus*, according as the $(n + 1)^{th}$ value of dy, counting from the point of contact, is *greater* or *less* than the nth value of dy. Conversely, if the $(n + 1)^{th}$ differential of y is *plus*, the $(n + 1)^{th}$ value of dy is greater than the nth; if *minus* the $(n + 1)^{th}$ value of dy is less than the nth.

II. CURVATURE.

Direction of Curvature.

35. Let KL be a curve, whose concavity is turned downward, that is, in the direction of negative ordinates; let AP, BQ, and CQ' be consecutive ordinates, AB and BC being equal to dx; prolong PQ toward T. Then will RQ, and $R'Q'$, be consecutive values of dy, and consequently their difference, $R'Q' - RQ$, will be the value of d^2y at P. The right angled triangles, PRQ, and $QR'S$, have their bases, and the angles at their bases, equal; hence, their altitudes, RQ, and $R'S$, are equal; but $R'Q'$ is less than $R'S$, because Q' is nearer $-\infty$ than S; hence, $R'Q'$ is less than RQ, and consequently the value of d^2y is *negative*. Conversely, if d^2y is *negative*, $R'Q'$ is less than $R'S$, and the curve in passing from P is *concave downward*. In like manner, it may be shown that the curve is *concave upward* at P, when d^2y is *positive*. Because the sign of the second differential coefficient is the same as that of the second differential, we have the following rule for determining the direction of curvature, in passing from any point toward the right.

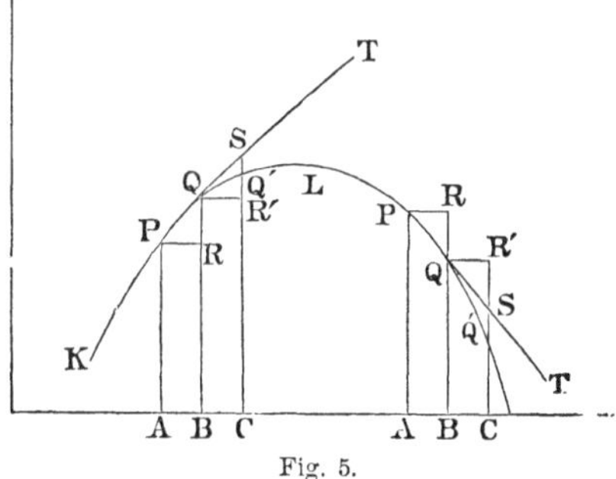

Fig. 5.

Find the second differential coefficient of the ordinate at the point; if this is negative, the curve is concave downward, if positive, it is concave upward.

We may regard y as the independent variable, and find the second differential coefficient of x; if this be negative at any point, the concavity is turned toward the left, if positive, toward the right.

Example.—Let it be required to determine the direction of curvature at any point of a parabola.

Assume the equation, $y^2 = 2px$; whence, by differentiation and reduction,

$$\frac{d^2y}{dx^2} = -\frac{p^2}{y^3}, \text{ and } \frac{d^2x}{dy^2} = \frac{1}{p}.$$

The *first* of these is negative for positive values of y, and positive for negative values of y; hence, the part of the curve above the axis of X is concave downward, and the part below is concave upward. The *second* is positive for all values of x and y; hence, the curve is everywhere concave toward the right.

If the tangent at P have a contact of the nth order, n being greater than 1, the second differential coefficient of y at that point will be 0 (Art. 34). In this case it may be shown, as above, that the curve bends downward in passing toward the right, when the $(n+1)^{\text{th}}$ value of dy, counting from P, is less than the nth value of dy, and upward when the $(n+1)^{\text{th}}$ value of dy is greater than the nth. The former condition is satisfied when the $(n+1)^{\text{th}}$ differential coefficient of y is *negative*, and the latter when it is *positive* (Art. 34).

Amount of Curvature.

36. The *change of direction* of a curve in passing from an element to the one next in order, is the angle between the second element and the prolongation of the first. This angle is called the *angle of contingence*, it is *negative* when the curve bends downward, and *positive* when it bends upward. The total change of direction in passing over any arc is equal to the algebraic sum of the angles of contingence at every point of that arc.

The *curvature* of a curve is the *rate* at which it changes direction. When the curvature is uniform, as in the circle, it is measured by the total change of direction in any arc, divided by the length of that arc. In any curve whatever the curvature may be regarded as uniform for an infinitely small arc; hence, the curvature at any point of a curve, is equal to the angle of contingence at that point, divided by the corresponding element of the curve. Denoting the angle of contingence by $d\theta$, the corresponding element by ds, and the curvature by c, we have,

$$c = \frac{d\theta}{ds} \quad \ldots\ldots \quad (1)$$

Osculatory Circle.

37. An *osculatory circle*, to a given curve, is a circle that has three consecutive points in common with that curve. Thus, if the circle whose centre is C, pass through the three consecutive points P, Q, and R, of the curve, KQ, it is osculatory to that curve at P.

The circle and curve have two consecutive tangents, PT and QT', and two consecutive normals, PC and QC, in common; they have also the angle of contingence at Q, in common; hence, they have the same curvature at the point P. Because PC and QC are perpendicular to PT, and QT', their included angle, is equal to

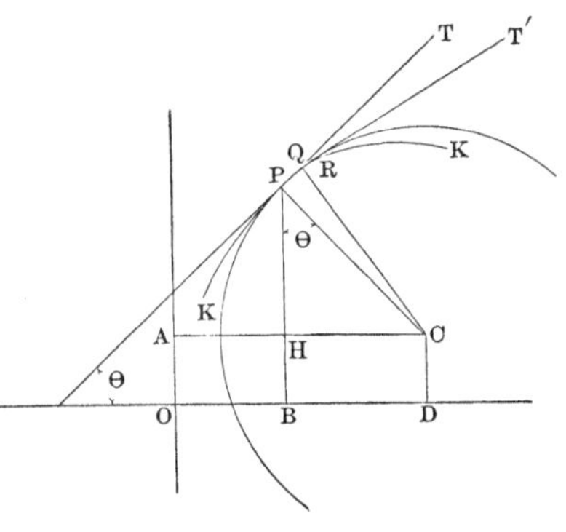

Fig. 6.

the angle of contingence, $d\theta$; hence, $PQ = CP \times d\theta$, or denoting PQ by ds and CP by R, we have, $ds = Rd\theta$, or,

$$\frac{d\theta}{ds} = \frac{1}{R} \quad . \; . \; . \; . \; . \quad (1)$$

This equation shows that the curvature of a curve at any point is equal to the reciprocal of the radius of the osculatory circle at that point; for this reason, the radius of the osculatory circle is called the *radius of curvature.*

Radius of Curvature.

38. If we denote the inclination of the tangent, PT, to the axis of X by θ, the angle of contingence, TQT', equal to PCQ, will be equal to $d\theta$; but we know that $\theta = \tan^{-1}\frac{dy}{dx}$; differentiating, we have,

$$d\theta = \frac{\frac{d^2y}{dx}}{1 + \frac{dy^2}{dx^2}} = \frac{dxd^2y}{dy^2 + dx^2}.$$

Finding the value of R from equation (1), Art. 37, and replacing ds by its value, from Art. 5, we have,

$$R = \frac{(dx^2 + dy^2)^{\frac{3}{2}}}{dxd^2y} \quad . \; . \; . \; . \; . \quad (1)$$

Dividing both terms of the second member by dx^3, and denoting the first differential coefficient of y by q', and its second differential coefficient by q'', we have,

$$R = \frac{(1 + q'^2)^{\frac{3}{2}}}{q''} \quad . \; . \; . \; . \; . \quad (2)$$

When the curve bends downward, $d\theta$ is negative, and consequently R is *negative;* when the curve bends upward, both are *positive.*

It is sometimes convenient to regard R, or some other quantity on which R depends, as the independent variable. In this case both dy and dx are variable, and we have,

$$d\theta = \frac{dxd^2y - dyd^2x}{dx^2 + dy^2};$$

and this in (1), Art. 37, gives,

$$R = \frac{(dx^2 + dy^2)^{\frac{3}{2}}}{dxd^2y - dyd^2x} \quad \ldots\ldots \quad (3)$$

Example.—Required the value of R for the parabola. Assume the equation, $y^2 = 2px$.

By differentiation and reduction, we have,

$$\frac{dy}{dx} = q' = \frac{p}{y}, \text{ and } \frac{d^2y}{dx^2} = q'' = -\frac{p^2}{y^3};$$

substituting these in (2), we have,

$$R = \left(1 + \frac{p^2}{y^2}\right)^{\frac{3}{2}} \times -\frac{y^3}{p^2} = -\frac{(y^2 + p^2)^{\frac{3}{2}}}{p^2}.$$

The value of R is least possible when $y = 0$, increases as y increases, and becomes ∞ when $y = \infty$. Hence, the curvature of the parabola is greatest at the vertex, and diminishes as the curve recedes from the vertex.

Co-ordinates of the Centre of the Osculatory Circle.

39. In the right angled triangle, PHC, PH is perpendicular to ED, and PC to EP; hence, the angle, HPC, is equal to θ. Denoting the abscissa, AC, by α, and the ordinate, DC, by β, we have,

$$\alpha = AH + HC, \text{ and } \beta = BP - HP \; . \; . \; . \; . \; . \quad (1)$$

AH and BP are the co-ordinates of P, denoted by x and y; HP is equal to $PC\cos\theta$, and HC to $PC\sin\theta$; substituting for PC, or R, its value found in the last article, remembering that it is *negative* in the case under consideration, (Art. 38), replacing $\sin\theta$, and $\cos\theta$, by their values taken from Art. 5, and reducing, we have,

$$\left.\begin{aligned} \alpha &= x - \frac{dx^2 + dy^2}{d^2y} \cdot \frac{dy}{dx} = x - \left(\frac{1 + q'^2}{q''}\right)q' \\ \beta &= y + \frac{dx^2 + dy^2}{d^2y} = y + \frac{1 + q'^2}{q''} \end{aligned}\right\} \; . \; . \; . \; . \; . \quad (2)$$

Locus of the Centre of Curvature.

40. A line drawn through the centres of the osculatory circles at every point of a curve is called the *evolute* of that curve, and the given curve, with respect to its evolute, is called an *involute*. If we consider any radius of curvature of a curve, we see that it intersects the one that precedes and the one that follows it, and the distance between the points of intersection is an element of the evolute; hence, the radius is normal to the involute, and tangent to the evolute. We also see that the difference between two consecutive radii of curvature is equal to the corresponding element of the evolute; hence, the difference between any two radii of curvature of a given curve is equal to the corresponding arc of the evolute.

The evolute of a curve may be constructed by drawing normals to the curve, and then drawing a curve tangent to them all: the closer the normals, the more accurate will be the construction. An involute may be constructed by wrapping a thread round the evolute, holding it tense, and then unwrapping it. Every point of the thread describes a curve, which is an involute of the given evolute. Practically, the evolute is cut out of a board, or other solid material. In this manner the outlines of the teeth of wheels are sometimes marked out, the circle being taken as an evolute.

Equation of the Evolute of a Curve.

41. The equation of the evolute of any curve may be found by combining the equation of the curve with formulas (2), Art. 39, and eliminating x and y. The method will be best illustrated by an example; thus, let it be required to find the equation of the evolute of the common parabola.

We have already found $q' = \frac{p}{y}$, and $q'' = -\frac{p^2}{y^3}$ (Art. 38); substituting these in (2), Art. 39, we have, after reduction,

$$\alpha = x + \frac{y^2 + p^2}{p} = 3x + p, \quad \therefore \quad x = \frac{1}{3}(\alpha - p).$$

$$\beta = y - \frac{y(y^2 + p^2)}{p^2} = -\frac{y^3}{p^2}, \quad \therefore \quad y = -\beta^{\frac{1}{3}}p^{\frac{2}{3}};$$

substituting the values of x and y in the equation $y^2 = 2px$, we have,

$$\beta^{\frac{2}{3}}p^{\frac{4}{3}} = \frac{2p}{3}(\alpha - p), \quad \therefore \beta^2 = \frac{8}{27p}(\alpha - p)^3,$$

which is the equation required. It is the equation of a semi-cubic parabola.

EXAMPLES.

1. Find the equation of the evolute of an ellipse

$$Ans.\ (a\alpha)^{\frac{2}{3}} + (b\beta)^{\frac{2}{3}} = (a^2 - b^2)^{\frac{2}{3}}$$

2. Find the equation of the evolute of an hyperbola.

$$Ans.\ (a\alpha)^{\frac{2}{3}} - (b\beta)^{\frac{2}{3}} = (a^2 + b^2)^{\frac{2}{3}}$$

3. Find the radius of curvature of an ellipse.

$$Ans.\ \rho = -\frac{(a^4y^2 + b^4x^2)^{\frac{3}{2}}}{a^4b^4}.$$

4. Find the radius of curvature of an equilateral hyperbola referred to its asymptotes, the equation being $xy = m$.

$$Ans.\ \rho = \frac{(m^2 + x^4)^{\frac{3}{2}}}{2mx^3}.$$

III. SINGULAR POINTS OF CURVES.

Definition of a Singular Point.

42. A SINGULAR POINT of a curve, is a point at which the curve presents some peculiarity not common to other points. The most remarkable of these are, *points of inflexion, cusps, multiple points, and conjugate points.*

Points of Inflexion.

43. A POINT OF INFLEXION is a point at which the curvature changes, from being concave downward to being concave upward, or the reverse.

Inasmuch as the direction of curvature is determined by the sign of the second differential coefficient of the ordinate, it follows that this sign must change from $-$ to $+$, or from $+$ to $-$, in passing a point of inflexion. But a quantity can only change sign by passing through 0, or ∞. Hence, at a point of inflexion the second differential coefficient of y must be either 0, or ∞. If the corresponding values of x and y are such, that for values immediately preceding and following them, the second differential coefficient has contrary signs, then will each set of such values correspond to a point of inflexion.

EXAMPLES.

1. To find the points of inflexion on the curve whose equation is $y = b + (x - a)^3$.

We find, $\frac{d^2y}{dx^2} = 6(x - a)$; this, put equal to 0, gives $x = a$, which, in the equation of the curve, gives $y = b$. When $x < a$, the second differential coefficient is *negative*, and when $x > a$, it is positive. Hence, the point whose co-ordinates are a and b, is a point of inflexion.

2. To find the points of inflexion on the curve whose equation is $y = b + \frac{1}{10}(x - a)^5$.

We find, $\frac{d^2y}{dx^2} = 2(x - a)^3$; this placed equal to 0, gives $x = a$, whence, $y = b$. When $x < a$, the second differential coefficient is *negative*, and when $x > a$, it is *positive*. Hence, the point whose co-ordinates are a and b, is a point of inflexion.

3. Find the co-ordinates of the points of inflexion on the curve whose equation is $y = 2r\sqrt{\frac{2r - x}{x}}$.

Ans. $x = \frac{3}{2}r$, and $y = \frac{2}{3}r\sqrt{3}$.

4. Find the co-ordinates of the points of inflexion on the line whose equation is $y = \frac{x^2(a^2 - x^2)}{a^3}$.

Ans. $x = \pm \frac{1}{6} a \sqrt{6}$, and $y = \frac{5}{36} a$.

Cusps.

44. A CUSP is a point at which two branches terminate, being tangential to each other. There are two species of cusps; the *ceratoid*, named from its resemblance to the horns of an animal, and the *ramphoid*, named from its resemblance to the beak of a bird. The first is shown at E, and the second at A.

The method of determining the position and nature of a cusp point will be best shown by examples.

1. Let us take the curve whose equation is $y = b \pm (x - a)^{\frac{3}{2}}$. From this we find,

$$\frac{dy}{dx} = \pm \frac{3}{2}(x - a)^{\frac{1}{2}}, \text{ and } \frac{d^2y}{dx^2} = \pm \frac{3}{4}(x - a)^{-\frac{1}{2}}.$$

For $x = a$, $\frac{dy}{dx} = 0$, and $\frac{d^2y}{dx^2} = \infty$. For $x < a$, both are imaginary, as is also the value of y. For $x > a$, both are real, and each has two values, one *plus* and the other *minus*. Hence, the point whose co-ordinates are a and b is a cusp of the first species.

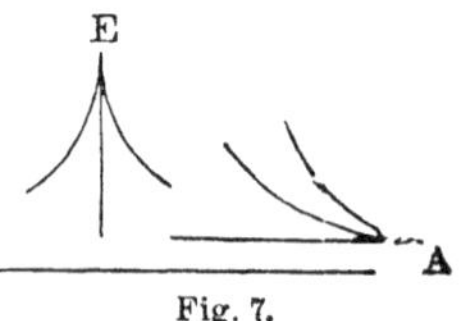

Fig. 7.

2. Let us take the curve whose equation is $y = x^2 \pm x^{\frac{5}{2}}$. From it we find,

$$\frac{dy}{dx} = 2x \pm \frac{5}{2} x^{\frac{3}{2}}, \text{ and } \frac{d^2y}{dx^2} = 2 \pm \frac{15}{4} x^{\frac{1}{2}}.$$

From the given equation we see that no part of the curve lies to the left of the axis of y, and that there are two infinite branches to the right of that axis.

The values of $\frac{dy}{dx}$ are both 0 at the origin; and at that point both values of $\frac{d^2y}{dx^2}$ are positive; hence, the origin is a cusp of the second species.

A discussion of the above equations shows that the upper branch has its concavity always upward; the second branch has a point of inflexion for, $x = \frac{64}{225}$; its slope is 0 for $x = \frac{16}{25}$; and it cuts the axis of x at the distance 1 from the origin; from the origin to the point of inflexion, it curves upward; after that point it curves downward. The shape of the curve is indicated in the figure.

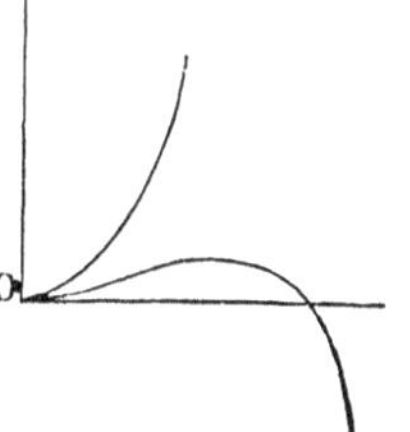

Fig. 8.

For the *ceratoid* the values of the second differential coefficient of y have contrary signs, for the *ramphoid* they have the same sign.

Multiple Points.

45. A MULTIPLE POINT is a point where two or more branches of a curve intersect, or touch each other.

If the branches intersect, $\frac{dy}{dx}$ will have as many values at that point as there are branches; if they are tangent, these values will be equal.

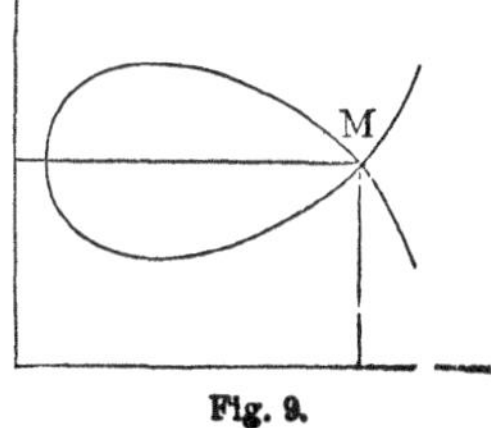

Fig. 9.

EXAMPLES.

1. Take the curve whose equation is $y^2 = x^2 - x^4$.

For every value of x there are two values of y equal with contrary signs; hence, the curve is symmetrically situated with respect to the axis of x. For $x = 0$, we have $y = \pm 0$; hence, both branches pass through the origin. From the equation, we find,

$$\frac{dy}{dx} = \frac{x}{y} - \frac{2x^3}{y} = \pm \frac{1}{\sqrt{1 - x^2}} \mp \frac{2x^2}{\sqrt{1 - x^2}}.$$

For $x = 0$, $\frac{dy}{dx} = \pm 1$, hence the origin is a multiple point, by intersection. The curve is limited in both directions.

2. Take the curve whose equation is $y = \pm \sqrt{x^5 + x^4}$.

The two branches are symmetrically placed with respect to the axis of x, both pass through the origin, forming a loop on the left and extending to infinity on the right. We find, from the equation of the curve,

$$\frac{dy}{dx} = \pm \frac{5x^4 + 4x^3}{2\sqrt{x^5 + x^4}} = \pm \frac{5x^2 + 4x}{2\sqrt{x + 1}}.$$

For $x = 0$, we have $\frac{dy}{dx} = \pm 0$; hence, the origin is a multiple point, the branches being tangent to each other and to the axis of x.

3. The curve whose equation is $y = \pm x\sqrt{\frac{a^2 - x^2}{a^2 + x^2}}$, has a multiple point at the origin, and is composed of two pointed loops, one on the right and the other on the left of the origin.

4. The curve whose equation is $y^2 = (x - 2)(x - 3)^2$ is symmetrical with respect to the axis of x and consists of a looped branch which extends from $x = 2$ to $x = 3$; at the latter point the upper part passes below the axis of x and lower part passes above that axis, forming a multiple point; beyond $x = 3$ the two parts continually diverge, extending to $x = \infty$.

Conjugate Points.

46. A CONJUGATE, or ISOLATED POINT, is a point whose co-ordinates satisfy the equation of a curve, but which has no consecutive point. Because it has no consecutive point, the value of the first differential coefficient of y at it, is imaginary.

EXAMPLES.

1. Take the curve whose equation is $y = \pm x\sqrt{x - a}$.

In this case $x = 0$ and $y = 0$, satisfy the equation, but no other value of x less than a gives a point of the curve. Furthermore, we have,

$$\frac{dy}{dx} = \pm \frac{1}{2} \frac{3x^2 - 2ax}{\sqrt{x^3 - ax^2}},$$

which for $x = 0$ is imaginary, as also for all values of $x < a$.

2. Take the curve whose equation is

$$y = ax^2 \pm \sqrt{x(1 - \cos x)}.$$

For every positive value of x, there are two values of y, and consequently two points, except when $\cos x = 1$, when the two points reduce to one. These points constitute a series of loops like the links of a chain, having for a diametral curve a parabola whose equation is $y = ax^2$. For every negative value of x the values of y are imaginary, except when $\cos x = 1$; in these cases we have a series of isolated points situated on the diametral curve, whose equation is $y = ax^2$. We have, by differentiating and reducing,

$$\frac{dy}{dx} = 2ax \pm \frac{\sqrt{1 - \cos x} + x\sqrt{1 + \cos x}}{2\sqrt{x}},$$

which is imaginary for negative values of x.

IV. Maxima and Minima.

Definitions of Maximum and Minimum.

47. A function of one variable is at a *maximum* state when it is *greater* than the states that immediately precede and follow it; it is at a *minimum* state, when it is *less* than the states that immediately precede and follow it. Thus, if KL be the curve of the function, BN will be a *maximum*, because it is greater than AM and CO, and EQ will be a *minimum*, because it is less than DP and FR.

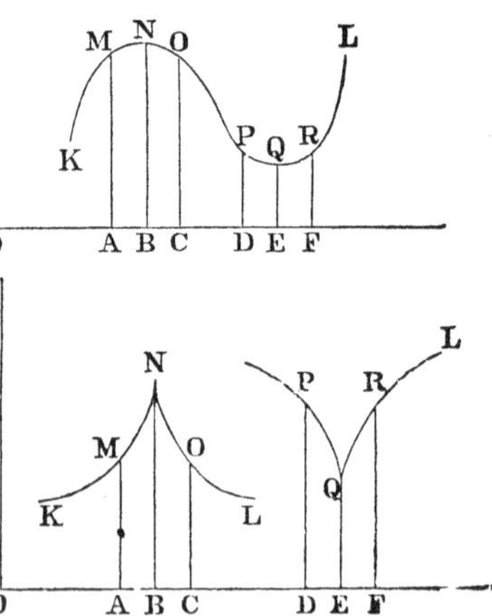

Fig. 10.

Analytical Characteristics of a Maximum or Minimum.

48. If we examine the figures given in the last article, we see that the slope of the curve KL, or the differential coefficient of the function, changes from $+$ to $-$ in passing a maximum, and from $-$ to $+$ in passing a minimum. But a varying quantity can only change sign by passing through 0, or ∞; hence, the first differential coefficient of the function must be either 0, or ∞, at either maximum or minimum state.

If, therefore, we find the first differential coefficient of the function, and set it equal to 0 *and* ∞, *we shall have two equations which will give all the values of the variable that belong to either a maximum or minimum state of the function.*

They may also give other values; hence the necessity of testing each root separately. This may be done by the following rule:

Subtract from, and add to, the root to be tested, an infinitely small quantity; substitute these successively in the first differential coefficient; if the first result is plus and the second minus, the root corresponds to a maximum; if the first is minus and the second plus, it corresponds to a minimum; if both have the same sign, it corresponds to neither a maximum nor a minimum.

The maximum or minimum value may be found by substituting the corresponding root in the given function.

EXAMPLES.

1. Let $y = 3 + (x - 2)^2$.

By the rule, we have, $\frac{dy}{dx} = 2(x - 2) = 0, \quad \therefore\ x = 2.$

Substituting for x the values, $2 - dx$, and $2 + dx$, we find for the corresponding values of the differential coefficient $-2dx$, and $+2dx$; hence, $x = 2$, corresponds to a *minimum*, which is $y = 3$.

2. Let $y = 4 - (x - 3)^{\frac{2}{3}}$.

We have,

$$\frac{dy}{dx} = -\frac{2}{3}(x - 3)^{-\frac{1}{3}} = \infty; \quad \therefore x = 3.$$

For $x = 3 - dx$, and $3 + dx$, we have the differential coefficient equal to

$$\frac{2}{3\sqrt[3]{dx}}, \text{ and } -\frac{2}{3\sqrt[3]{dx}},$$

hence, $x = 3$ corresponds to a maximum, $y = 4$.

3. Let $y = 3 + 2(x - 1)^3$.

We have,

$$\frac{dy}{dx} = 6(x - 1)^2 = 0; \quad \therefore x = 1.$$

Substituting $1 - dx$, and $1 + dx$, for x, in the first differential coefficient, we have in both cases a positive result; hence, $x = 1$ does not correspond to either a maximum, or minimum.

The preceding method is applicable in all cases; but, when the first differential coefficient of the function is 0, as it is in most instances, there is an easier process for testing the roots. In this case, the function being represented by the ordinate of a curve, the maximum and minimum states correspond to points at which the tangent is horizontal; furthermore, the curve lies wholly *above*, or wholly *below*, the tangent at the point of contact, that is, the tangent does not cut the curve at that point; hence, the contact is of an odd order, and consequently *the first of the successive differential coefficients of the function that does not reduce to* 0 *must be of an even order* (Art. 34). If this is *negative*, the curve bends downward after passing the point of contact, and the root corresponds to a *maximum*, if *positive* the curve bends upward, and the root corresponds to a *minimum* (Art. 35). Hence the following practical rule for finding the values of the variables that correspond to maxima and minima:

Place the first differential coefficient of the function equal to 0, *and solve the resulting equation; substitute each root in the successive differential coefficients of the function, until one is found that does not reduce to* 0; *if this is of an even order and negative, the root corresponds to a maximum, if of an even order and positive, to a minimum; but if of an odd order, it corresponds to neither maximum nor minimum.*

EXAMPLES.

1. Let $y = x^2 - 3x + 2$.

We have,

$$\frac{dy}{dx} = 2x - 3 = 0; \quad \therefore \ x = \frac{3}{2}.$$

Also,

$$\frac{d^2y}{dx^2} = 2; \quad \therefore \ \left[\frac{d^2y}{dx^2}\right]_{\frac{3}{2}} = +\,2.$$

Hence, $x = \frac{3}{2}$, corresponds to a minimum state, which is,

$$y = [x^2 - 3x + 2]_{\frac{3}{2}} = -\frac{1}{4}.$$

The symbol $\left[\frac{d^2y}{dx^2}\right]_{\frac{3}{2}}$ is used to denote what $\frac{d^2y}{dx^2}$ becomes when the variable that enters it is made equal to $\frac{3}{2}$.

2. $y = 4 - (x - 3)^4$.

$$\frac{dy}{dx} = - 4(x - 3)^3 = 0; \quad \therefore \ x = \mathbf{3};$$

$$\frac{d^2y}{dx^2} = - 12(x - 3)^2; \qquad \therefore \ \left[\frac{d^2y}{dx^2}\right]_3 = \mathbf{0};$$

we also find,

$$\left[\frac{d^3y}{dx^3}\right]_3 = 0; \text{ and } \left[\frac{d^4y}{dx^4}\right]_3 = - 24.$$

Hence, $[y]_3 = 4$, is a maximum.

In applying the above rule, a *positive constant factor* may be suppressed at any stage of the process; for it is obvious that such suppression can in no way alter the *value* of the roots to be found, or the *signs* of the successive differential coefficients.

3. Let $y = 3x^3 - 9x^2 - 27x + 30$.

Rejecting the factor, + 3, and denoting the result by u, we have,

$$\frac{du}{dx} = 3x^2 - 6x - 9 = 3(x^2 - 2x - 3) = 0; \ \therefore \ x = 3, x = - 1.$$

Rejecting the factor, + 3, and denoting the resulting value of the function by u', we have,

$$\left[\frac{d^2u'}{dx^2}\right]_3 = 4, \text{ and } \left[\frac{d^2u'}{dx^2}\right]_{-1} = - 4.$$

The first corresponds to a minimum and the second to a maximum. Substituting in the given function, we have, for the maximum and minimum values,

$$y' = 45, \text{ and } y'' = - 51.$$

When the first differential coefficient is composed of

variable factors, the method of finding the corresponding values of the second differential coefficient may be simplified as follows: let us have,

$$\frac{dy}{dx} = P \times Q;$$

P and Q being functions of x, and suppose that $[P]_a = 0$. By the rule for differentiating, we have,

$$\frac{d^2y}{dx^2} = P\frac{dQ}{dx} + Q\frac{dP}{dx};$$

but because P becomes 0 when $x = a$, we have,

$$\left[\frac{d^2y}{dx^2}\right]_a = \left[Q\frac{dP}{dx}\right]_a.$$

Hence, *find the differential coefficient of the factor that reduces to* 0, *multiply it by the other factor, and substitute the root in the product. The result is the same as that found by substituting the root in the second differential coefficient.*

4. Let $y = (x - 3)^2(x - 2)$.

We have,

$$\frac{dy}{dx} = 2(x - 3)(x - 2) + (x - 3)^2 = (x - 3)(3x - 7) = 0;$$

$$\therefore\ x = 3,\ x = \frac{7}{3}.$$

By the principle just demonstrated,

$$\left[\frac{d^2y}{dx^2}\right]_3 = [3x - 7]_3 = +2; \quad \left[\frac{d^2y}{dx^2}\right]_{\frac{7}{3}} = [3(x - 3)]_{\frac{7}{3}} = -2.$$

Hence, $[y]_3 = 0$, is a minimum, and $[y]_{\frac{7}{3}} = \frac{4}{27}$, is a maximum.

If y represents any function of x, and if $u = y^n$, we have,

$$\frac{du}{dx} = ny^{n-1}\frac{dy}{dx}.$$

If $x = a$ reduces $\frac{dy}{dx}$ to 0, but does not reduce y^{n-1} to 0, we have, from what precedes,

$$\left[\frac{d^2u}{dx^2}\right]_a = \left[ny^{n-1}\frac{d^2y}{dx^2}\right]_a.$$

If ny^{n-1} is $+$, any value that makes y a *maximum*, or *minimum*, also makes u a *maximum*, or *minimum*; if ny^{n-1} is $-$, any value that makes y a *maximum*, or *minimum*, makes u a *minimum*, or *maximum*. The converse is also true, if we except the values that make y^{n-1} equal to 0. Hence, if care be taken to reject the exceptional values, we may throw off a radical sign in seeking for a maximum or minimum.

5. Let $y = \sqrt{4a^2x^2 - 2ax^3}$.

Throwing off the radical sign, suppressing the fact $+ 2a$, and denoting the result by u, we have,

$$\frac{du}{dx} = 4ax - 3x^2 = 0; \;\therefore\; x = 0,\; x = \frac{4a}{3}.$$

The value, $x = 0$, makes $y = 0$, and is to be rejected. The value $x = \frac{4}{3}a$ makes $\frac{d^2u}{dx^2}$ *negative*, and gives $y = \frac{8a^2}{3\sqrt{3}}$, a *maximum*.

In like manner, if we have $u = ly$, we have,

$\frac{du}{dx} = \frac{1}{y} \cdot \frac{dy}{dx}$, and if $\left[\frac{dy}{dx}\right]_a = 0$, we have, as before,

$$\left[\frac{d^2u}{dx^2}\right]_a = \left[\frac{1}{y} \cdot \frac{d^2y}{dx^2}\right]_a.$$

Hence, we infer, as before, that we may, with proper precaution, treat the logarithm of a function instead of the function itself, and the reverse.

6. Let $y = \frac{x^2 - 3x + 2}{x^2 + 3x + 2}$, or $y = \frac{(x-1)(x-2)}{(x+1)(x+2)}$.

Passing to logarithms, and denoting ly by u, we have,

$$u = l(x-1) + l(x-2) - l(x+1) - l(x+2).$$

Whence,

$$\frac{du}{dx} = \frac{1}{x-1} + \frac{1}{x-2} - \frac{1}{x+1} - \frac{1}{x+2} = 0,$$

or,
$$\frac{6(x^2-2)}{x^4 - 5x^2 + 4} = 0; \quad \therefore x = \pm\sqrt{2}.$$

Both values of x make y negative. The first makes $\frac{d^2u}{dx^2}$ *negative*, and the second makes it *positive*. Hence we have, $y = 12\sqrt{2} - 17$, *minimum*, and $y = -12\sqrt{2} - 17$, *maximum*.

PROBLEMS IN MAXIMA AND MINIMA.

1. Divide 21 into two parts, such that the less multiplied by the square of the greater, shall be a maximum.

Solution.—Let x be the greater, and $21 - x$ the less. We have, for the equation of the problem, $y = (21 - x)x^2$. By the rule we find that $x = 14$ makes y a maximum Hence, the parts are 14 and 7.

2. Find a cylinder whose total surface is equal to S, and whose volume is a maximum.

Solution.—Denote the radius of the base by x, the altitude by y, and the volume by V. From the geometrical relations of the parts, we have,

$$S = 2\pi x^2 + 2\pi x \times y, \text{ and } V = \pi x^2 \times y.$$

Combining, we have, for the equation of the problem,

$$V = \pi x^2 \frac{S - 2\pi x^2}{2\pi x} = \frac{1}{2}(Sx - 2\pi x^3).$$

V is a maximum when $x = \sqrt{\frac{S}{6\pi}}$; whence, $y = 2\sqrt{\frac{S}{6\pi}}$, or $y = 2x$. That is, the altitude is equal to twice the radius of the base.

3. Find the maximum rectangle that can be inscribed in an acute-angled triangle whose base is 14 feet, and altitude 10 feet.

Ans. The base is 7 feet, and the altitude 5 feet.

4. Find the maximum triangle that can be constructed on a given base, and having a given perimeter.

Solution.—Denote the base by b, a second side by x, the third side by $2p - b - x$, the perimeter by $2p$, and the area by A. From the formula for the area in terms of the three sides, we find the equation of the problem,

$$A = \sqrt{p(p - b)(p - x)(b + x - p)}.$$

For $x = p - \frac{1}{2}b$, A is a maximum. Hence, the triangle is isosceles.

5. To find the maximum isosceles triangle that can be inscribed in a circle.

Ans. An equilateral triangle.

6. To find the minimum isosceles triangle that can be circumscribed about a circle.

Ans. An equilateral triangle.

7. To find the maximum cone that can be cut from a sphere whose radius is r.

Ans. The radius of the base is $\frac{2r\sqrt{2}}{3}$, and the altitude is $\frac{4}{3}r$.

8. To find the maximum cylinder that can be cut from a cone whose altitude is h, and the radius of whose base is r.

Ans. The radius of its base is $\frac{2}{3}r$, and its altitude $\frac{1}{3}h$.

9. To find the altitude of the maximum cylinder that can be cut from a sphere whose radius is r.

Ans. Altitude $= \frac{2}{3}r\sqrt{3}$.

10. To find the maximum rectangle that can be inscribed in an ellipse whose semi-axes are a and b.

Ans. The base is $a\sqrt{2}$, and the altitude $b\sqrt{2}$.

11. To find the maximum segment of a parabola that can be cut from a right cone whose altitude is h, and the radius of whose base is r.

Ans. The axis is equal to $\frac{3}{4}\sqrt{h^2 + r^2}$.

12. To find the maximum parabola that can be inscribed in a given isosceles triangle.

Ans. The axis is equal to $\frac{3}{4}$ths the altitude.

13. To find the maximum cone whose surface is constant, and equal to S.

Ans. The radius of the base is $\frac{1}{2}\sqrt{\frac{S}{\pi}}$.

14. To find the maximum cylinder that can be cut from a given oblate spheroid, whose semi-axes are a and b.

Ans. The radius of the base $= a\sqrt{\frac{2}{3}}$,

and the altitude $= b\frac{2}{\sqrt{3}}$.

15. From the corners of a rectangle whose sides are a and b, four squares are cut and the edges turned up to form a rectangular box. Required the side of each square when the box holds a maximum quantity.

Ans. $\frac{a+b}{6} - \frac{1}{6}\sqrt{a^2 - ab + b^2}$.

16. To find the minimum parabola that will circumscribe a given circle.

Solution.—Let x and y be the co-ordinates of the point of contact, a and b the terminal co-ordinates, $2p$ the varying parameter, and r the radius of the circle. For any circumscribed parabola, we have,

$$a = x + p + r, \text{ and } b = \sqrt{2pa} = \sqrt{2p(x + p + r)} \ldots\ldots (1)$$

But from the triangle, whose sides are the radius, the subnormal, and the ordinate of the point of contact, we have,

$$y^2 = r^2 - p^2 = 2px; \quad \therefore x = \frac{r^2 - p^2}{2p}, \; a = \frac{(p+r)^2}{2p},$$

$$\text{and } b = p + r \ldots\ldots (2)$$

It will be shown hereafter that the area of a parabola is two-thirds the rectangle having the same base and altitude.

Assuming this property, and denoting the area by A, we have, for the equation of the problem,

$$A = \frac{2}{3} \cdot 2a \cdot b = \frac{2}{3}\frac{(p+r)^3}{p} \quad \ldots\ldots \quad (3)$$

Applying the rule, and remembering that p and A are variable, we find that A is a minimum when $2p = r$.

17. Find the maximum difference between the sine and versed-sine of a varying angle.

Ans. When the arc is 45°.

18. Find the maximum value of y in the equation $y = x^{1-lx}$.

Ans. $y' = e^{\frac{1}{4}}$.

19. Divide the number 36 into two factors such that the sum of their squares shall be a minimum.

Ans. Each factor is equal to 6.

20. Divide a number m into such a number of equal parts that their continued product shall be a maximum.

Solution.—Let x denote the number of parts, $\frac{m}{x}$ one part, and $\left(\frac{m}{x}\right)^x$ the continued product; hence, the equation of the problem is,

$$y = \left(\frac{m}{x}\right)^x; \text{ or, } ly = xl\frac{m}{x} = xlm - xlx.$$

$$\frac{dy}{dx} = y(lm - 1 - lx) = 0; \quad \therefore\ lx = lm - 1 = lm - le = l\frac{m}{e},$$

$$\therefore\ x = \frac{m}{e}, \text{ and } y = e^{\frac{m}{e}}$$

$$\left[\frac{d^2y}{dx^2}\right]_{\frac{m}{e}} = \left[y \times -\frac{1}{x}\right]_{\frac{m}{e}} = -e^{\frac{m}{e}} \times \frac{e}{m}; \therefore y' = e^{\frac{m}{e}},$$

is a maximum.

21. What value of x will make the expression $\frac{\sin x}{1 + \tan x}$, a maximum?

Ans. $x = 45°$.

22. Find the fraction that exceeds its square by the greatest possible quantity.

Ans. $\frac{1}{2}$.

23. The illuminating power of a ray of light falling obliquely on a plane varies inversely as the square of the distance from the source, and directly as the sine of its inclination to the plane.

How far above the centre of a horizontal circle must a light be placed that the illumination of the circumference may be a maximum?

Ans. $\pm\frac{r}{\sqrt{2}}$.

Maxima and Minima of a Function of Two Variables.

49. A function of two variables is at its *maximum* state when it is greater, and at its *minimum* state when it is less than all its consecutive states.

If two parallel sections (Art. 23) be taken through a maximum or minimum ordinate of a surface, the ordinate will be a maximum or minimum in each section, and in nearly every practical case the reverse will hold true. The exceptional cases will be those in which the maximum or minimum ordinate corresponds to a singular point of the surface. Omitting these, the problem is reduced to finding an ordinate that shall be a maximum, or a minimum in both the parallel sections through it This requires that $\frac{dz}{dy}$ and $\frac{dz}{dx}$ be simultaneously equal to 0. The equations thus obtained will determine all the values of x and y that correspond to maxima or minima states, and each set of such values may be tested in the manner explained in the last article, observing that for the section parallel to the plane xz, the successive differential coefficients of z, are found by supposing y constant, and for the section parallel to yz, they are found by supposing x constant.

EXAMPLES.

1. Let $z = x^3 - 3xy + y^3$.

We have,

$$\frac{dz}{dx} = 3x^2 - 3y = 0, \text{ and } \frac{dz}{dy} = 3y^2 - 3x = 0,$$

$$\therefore x = 0, y = 0; \text{ and } x = 1, y = 1.$$

Also,

$$\frac{d^2z}{dx^2} = 6x, \frac{d^3z}{dx^3} = 6; \text{ and } \frac{d^2z}{dy^2} = 6y, \frac{d^3z}{dy^3} = 6.$$

The first set of values reduce the second differential coefficients to 0, but not the third differential coefficients; hence, they are to be rejected. The second set of values make both of the second differential coefficients positive; hence, they correspond to a minimum, which is $z = -1$.

2. Let $z = x^3y^2 - x^4y^2 - x^3y^3$.

For, $x = \frac{1}{2}, y = \frac{1}{3}, \quad z = \frac{1}{432}$, a maximum.

3. Let $z = \sqrt{p(p-x)(p-y)(x+y-p)}$.

Dropping the radical sign, passing to logarithms, and denoting the logarithm of z^2 by u, we have,

$$u = l(p) + l(p-x) + l(p-y) + l(x+y-p).$$

Hence,

$$\frac{du}{dx} = -\frac{1}{p-x} + \frac{1}{x+y-p} = 0;$$

$$\frac{du}{dy} = -\frac{1}{p-y} + \frac{1}{x+y-p} = 0,$$

$$x = \frac{2}{3}p,\ y = \frac{2}{3}p, \text{ make } z \text{ a maximum.}$$

V. Singular Values of Functions.

Definition and Method of Evaluation.

50. A *singular value* of a function is one that appears under an indeterminate form, for a particular value of the variable. Thus, the expression $\frac{x - \sin x}{x^3}$ reduces to $\frac{0}{0}$, for the particular value, $x = 0$, whereas its real value is $\frac{1}{6}$.

Singular values that take the above form of indetermination, may often be detected, and their real value found by means of the calculus. Let $u = \frac{z}{y}$, in which z and y are functions of x that reduce to 0 for $x = a$. Clearing of fractions and differentiating, we have,

$$udy + ydu = dz;$$

if we make $x = a$, the second term disappears, and we have,

$$[u]_a = \left[\frac{dz}{dy}\right]_a.$$

If both dz and dy reduce to 0 for $x = a$, we have, in like manner,

$$[u]_a = \left[\frac{d^2z}{d^2y}\right]_a,$$

and so on. Hence, the following rule:

Divide the successive differentials of the numerator by the corresponding differentials of the denominator; substitute the particular value of the variable in the resulting fraction, continuing the operation till one is reached that is not indeterminate; the value thus found is the value required.

Thus, in the example above, we have,

$$\left[\frac{x - \sin x}{x^3}\right]_0 = \left[\frac{1 - \cos x}{3x^2}\right]_0 = \left[\frac{\sin x}{6x}\right]_0 = \left[\frac{\cos x}{6}\right]_0 = \frac{1}{6}.$$

EXAMPLES.

1. Find the value of $\left[\dfrac{x^3 + 2x^2 - x - 2}{x^3 - 1}\right]_1$. *Ans.* 2.

2. Find the value of $\left[\dfrac{x^2 - 2x}{x^4 - 2x^3 + 8x - 16}\right]_2$. *Ans.* $\frac{1}{8}$.

3. Find the value of $\left[\dfrac{x^3 - a^2x - ax^2 + a^3}{x^2 - a^2}\right]_a$. *Ans.* 0.

4. Find the value of $\left[\dfrac{1 - x^n}{1 - x}\right]_1$. *Ans.* n.

5. Find the value of $\left[\dfrac{e^x - e^{-x}}{l(1 + x)}\right]_0$. *Ans.* 2.

6. Find the value of $\left[\dfrac{1 - \cos x}{x}\right]_0$. *Ans.* 0.

7. Find the value of $\left[\dfrac{x^3 - 1}{x^3 + 2x^2 - x - 2}\right]_1$. *Ans.* $\frac{1}{2}$.

8. Find the value of $\left[\dfrac{x^4 - 1}{x^2 - 1}\right]_1$. *Ans.* 2.

9. Find the value of $\left[\frac{x^2 \tan x}{1 + \tan x}\right]_{\frac{\pi}{2}}$ *Ans.* $\frac{\pi^2}{4}$.

10. Find the value of $\left[\frac{x - \sqrt{2x^2 - a^2}}{2x - \sqrt{5x^2 - a^2}}\right]_a$. *Ans.* 2.

Singular values that appear under the forms $0 \times \infty$, $\frac{\infty}{\infty}$, and $\infty - \infty$, can be reduced to the form discussed, and then treated by the rule. The method of proceeding in each case will be illustrated by an example.

11. The expression $(1 - x) \tan \frac{\pi x}{2}$, which reduces to $0 \times \infty$, when $x = 1$, can be placed under the form $\frac{1 - x}{\cot \frac{\pi}{2} x}$, which, by the rule, reduces to $\frac{2}{\pi}$ for the particular value, $x = 1$.

12. The expression, $\tan \frac{\pi x}{2} \div \frac{x^2}{(x^2 - 1)}$, which reduces to $\frac{\infty}{\infty}$ for $x = 1$, can be placed under the form $\frac{x^2 - 1}{x^2 \cot \frac{\pi}{2} x}$, and this, by the rule, gives for $x = 1$, the value $-\frac{4}{\pi}$.

13. The expression, $x \tan x - \frac{\pi}{2} \sec x$, which reduces to $\infty - \infty$, for the value $x = \frac{\pi}{2}$, may be written $\frac{x \sin x - \frac{1}{2}\pi}{\cos x}$, which, by the rule, gives for $x = \frac{\pi}{2}$, the value -1.

VI. Elements of Geometrical Magnitudes.

Differentials of Lines, Surfaces, and Volumes.

51. Let AP and BQ be two consecutive ordinates of the curve, KL, whose equation is $y = f(x)$; then will PQ be the differential of the length of the curve, and $APQB$ will be the differential of the area, bounded by the curve, the axis of x, and any two ordinates; if we suppose the figure to revolve about the axis of x, the curve will generate a surface of revolution, and the area between the curve and axis will generate a volume of revolution; the surface generated by PQ, is the differential of the surface of revolution, and the volume generated by $APBQ$, is the differential of the volume of revolution. When the equation of KL is given, the values of these differentials may always be found in terms of x and dx, or of y and dy.

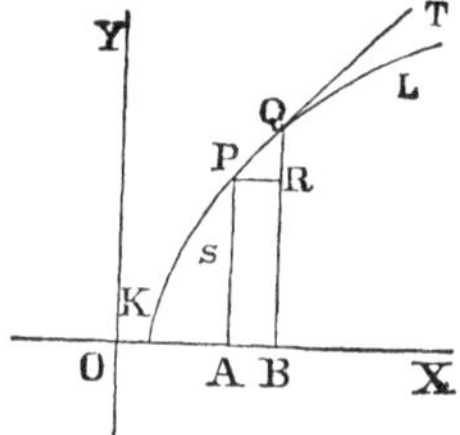

Fig. 11.

1°. Denote PQ by dL; we shall have, as in Art. 5,

$$dL = \sqrt{dx^2 + dy^2} \quad . \; . \; . \; . \; . \quad (1)$$

To apply this formula, we differentiate the equation of the curve, combine the resulting equation with that of the curve, so as to find the differential of one variable in terms of the other variable and its differential, and substitute this in the formula. Thus, let it be required to find the differential of the arc of a parabola. Assuming the equation of the parabola, $y^2 = 2px$, we find, by differentiation and combination,

$$dy = dx\sqrt{\frac{p}{2x}}, \text{ and } dx = \frac{y}{p}dy, \text{ or } dy^2 = dx^2\frac{p}{2x},$$

$$\text{and } dx^2 = \frac{y^2}{p^2}dy^2.$$

Substituting, and reducing, we have,

$$dL = dx\sqrt{\frac{2x + p}{2x}}, \text{ or } dL = \frac{dy}{p}\sqrt{p^2 + y^2}.$$

2°. Denote $APQB$ by dA; this is made up of two parts, the rectangle, $AR = ydx$, and the triangle, $RPQ = \frac{1}{2}dydx$; but the latter is an infinitesimal of the second order, it may be therefore neglected in comparison with the former; hence, we have,

$$dA = ydx \quad . \; . \; . \; . \; . \quad (2)$$

This formula may be applied in a manner similar to that just explained. Thus, if it be required to find the differential of the area of a parabola, we have,

$$dA = \frac{y^2}{p}dy; \text{ or, } dA = (\sqrt{2px})dx.$$

3°. Denote the surface generated by PQ by dS; this surface is that of a frustum of a cone, in which the radius of the upper base is y, the radius of the lower base, $y + dy$, and the slant height, $\sqrt{dx^2 + dy^2}$. Hence,

$$dS = \frac{1}{2}[2\pi y + 2\pi(y + dy)] \times \sqrt{dx^2 + dy^2}.$$

Neglecting dy in comparison with y, and reducing, we have,

$$dS = 2\pi y\sqrt{dx^2 + dy^2} \quad . \; . \; . \; . \; . \quad (3)$$

To find the differential of a paraboloid, we proceed as before. Substituting the values already found, and reducing, we have,

$$dS = 2\pi dx\sqrt{2px + p^2}\ ;\ \text{or,}\ dS = \frac{2\pi y dy}{p}\sqrt{p^2 + y^2}.$$

4°. Denote the volume generated by $APQB$, by dV; this volume is that of a frustum of a cone, in which the radius of the upper base is y, of the lower base, $y + dy$, and the altitude dx. Hence,

$$dV = \frac{[\pi y^2 + \pi(y + dy)^2 + \pi y(y + dy)]dx}{3}\ ;$$

neglecting dy in comparison with y, we have,

$$dV = \pi y^2 dx\ .\ .\ .\ .\ .\ (4)$$

Applying the formula to the paraboloid of revolution, we have, as before,

$$dV = \pi\frac{y^3}{p}dy,\ \text{or,}\ dV = 2\pi pxdx.$$

VII. Application to Polar Co-ordinates.

General Notions, and Definitions.

52. In a polar system, the radius vector is usually taken as the *function*, the angle being the *independent variable*. Let P be any point of the curve, PL, in the plane of the axes, OX, OY; let O be the *pole*, and OX the *initial line*, of a system of polar co-ordinates. Then will OP,

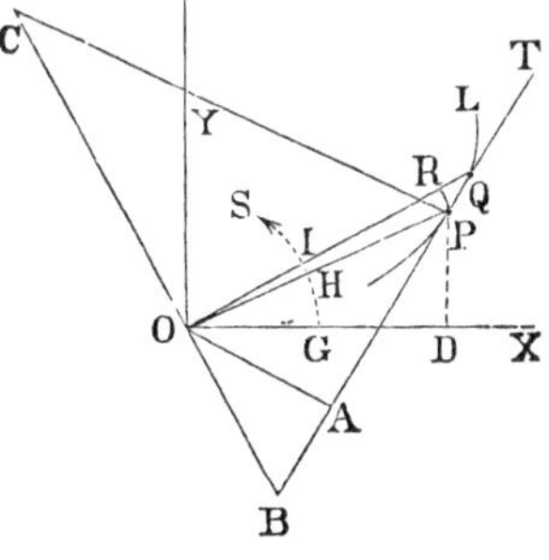

Fig. 12.

denoted by r, and XOP, denoted by φ, be the polar co-ordinates of P, and the equation of PL may be written under the form,

$$r = f(\varphi).$$

It will be convenient to express the values of φ in terms of π as a unit; in this case, it is laid off on the *directing circle*, GS, in the direction of the arrow when positive, and in a contrary direction when negative. Let GH be the measure of φ for the point P, and let HI be equal to the constant infinitesimal, $d\varphi$; draw OIQ, and with O as a centre describe the arc, PR; then will OP and OQ be consecutive radius vectors, P and Q will be consecutive points, RQ will be the differential of r, PQ will be the differential of the arc, POQ will be the differential of the area swept over by the radius vector, and PQ prolonged, will be tangent to the curve at P.

The line, BC, perpendicular to the radius vector, OP, is the *movable axis*, and the perpendicular distance, OA, is the ***polar** distance of the tangent.* Prolong the tangent to meet the movable axis at B; at P draw a normal, and produce it to meet the movable axis at C; then is OB the subtangent, PB the tangent, OC the subnormal, and PC the normal at the point P.

Useful Formulas.

53. 1°. *Differential of the arc.* Denote the differential of the arc, PQ, by dL; RP being infinitesimal may be regarded as a straight line perpendicular to OQ; it is equal to $rd\varphi$; hence,

$$dL = \sqrt{dr^2 + r^2 d\varphi^2} \quad . \ . \ . \ . \ . \ (1)$$

2°. *Differential of the area.* Denote the differential of the area swept over by the radius vector by dA; this is made up of the sector, OPR, and the triangle, RPQ; but, RPQ is an infinitesimal of the second order, and the sector is an infinitesimal of the first order; neglecting the former in comparison with the latter, and remembering that the area of a sector is equal to half the product of its arc and radius, we have,

$$dA = \frac{1}{2} r^2 d\varphi \quad . \quad . \quad . \quad . \quad . \quad (2)$$

3°. *Angle between the radius vector and tangent.* Denote the required angle by V; the angle, RQP, differs from the required angle, OPB, by an infinitesimal, and may, therefore, be taken for it; but *tan* RQP, equals RP, divided by RQ; hence, we have,

$$\tan V = \frac{r d\varphi}{dr} \quad . \quad . \quad . \quad . \quad . \quad (3).$$

4°. *Polar distance of the tangent.* Denote the distance OA by p; the triangles, QPR and QOA, are similar: hence,

$$QP : RP :: QO : OA.$$

But, PO differs from QO by an infinitesimal, and may, therefore, be taken for it; making this change, and substituting for the quantities their values, we have,

$$\sqrt{dr^2 + r^2 d\varphi^2} : r d\varphi :: r : p;$$

hence,

$$p = \frac{r^2 d\varphi}{\sqrt{dr^2 + r^2 d\varphi^2}} . \qquad (4)$$

5°. *Formula for subtangent.* Denote the subtangent by $S.T$; in the triangle, POB, the perpendicular, OB, is equal to the base, OP, multiplied by the tangent of the angle at the base; hence,

$$S.T = \frac{r^2 d\varphi}{dr} \quad \ldots\ldots \quad (5)$$

6°. *Formula for subnormal.* Denote the subnormal by $S.N$; the triangles, OPB and OCP, are similar; hence, $\tan OCP = \tan OPB$; but, OC, equals OP, divided by $\tan OCP$; hence,

$$S.N = \frac{dr}{d\varphi} \quad \ldots\ldots \quad (6)$$

7°. *Formula for the radius of curvature.* Denote the radius of curvature by ρ; let P and Q be two consecutive points of the curve, KL, and let PC and QC be normals at these points, meeting at C; then will C be the centre of the osculatory circle; draw OM perpendicular to PC, it will be parallel to the tangent at P, and, consequently, the intercept, PM, will be equal to the polar distance of the tangent denoted by p; draw OC, OP, and OQ. From the figure, we have,

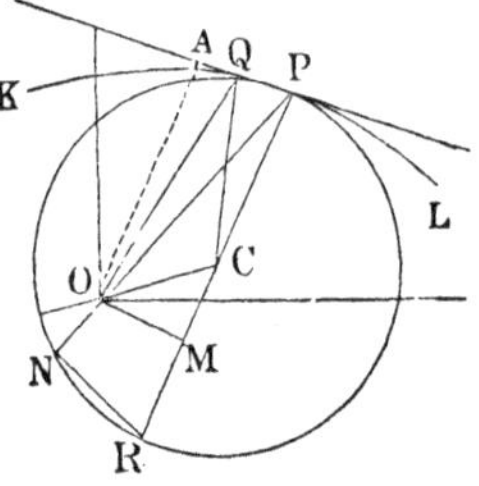

Fig. 13.

$$\overline{OC}^2 = r^2 + \rho^2 - 2p\rho.$$

If we pass from P to Q, r will become $r + dr$, p will become $p + dp$, and OC and ρ will remain unchanged. If, therefore, we differentiate the preceding equation under the supposition that r and p are variable, the resulting equation will express a relation between ρ, r, and p. Differentiating, we have,

$$2rdr - 2\rho dp = 0; \quad \therefore \rho = \frac{rdr}{dp} \quad . \; . \; . \; . \; . \quad (7)$$

8°. *Formula for the chord of curvature.* The chord of curvature is the chord of the osculatory circle that passes through the pole and the point of osculation. Denote it by C; prolong PC and PO till they meet the circle at R and N, and draw RN. The right angled triangles, PNR and PMO, are similar, hence,

$$PN : PR :: PM : PO,$$

or,

$$C : 2\rho :: p : r;$$

hence,

$$C = \frac{2p\rho}{r}, \text{ or } C = 2p\frac{dr}{dp} \quad . \; . \; . \; . \; . \quad (8)$$

Spirals.

54. If a straight line revolve uniformly about one of its points as a centre, and if, at the same time, a second point travel along the line, in accordance with any law, the latter point will describe a *spiral.* The part described in one revolution of the straight line is a *spire.* The fixed point is the *pole.* If we denote the distance from the pole to the generating point by r, and the corresponding angle, counted from the initial line, by φ, we have, for the general equation of spirals,

$$r = f(\varphi) \quad . \; . \; . \; . \; . \quad (1)$$

When this relation is algebraic, the spiral is said to be *algebraic;* when this relation is transcendontal, the spiral is said to be *transcendental.*

Among algebraic spirals, the most important are the *spiral of Archimedes*, the *parabolic spiral*, and the *hyperbolic spiral*, corresponding, respectively, to the *right line*, the *parabola*, and the *hyperbola*.

Spiral of Archimedes.

55. The equation of this curve is,

$$r = a\varphi \quad . \; . \; . \; . \; . \quad (2)$$

We see that the generating point is at the pole when the variable angle is 0, and that the radius vector increases uniformly with the variable angle, being always equal to a times the arc of the directing circle that has been swept over. Differentiating equation (2), we have,

$$dr = ad\varphi.$$

Substituting, in formulas (1) to (8), Art. 53, we find,

$$dL = ad\varphi\sqrt{1+\varphi^2}\,; \qquad S.T = a\varphi^2\,;$$

$$dA = \frac{1}{2}a^2\varphi^2 d\varphi\,; \qquad S.N = a\,;$$

$$\text{Tan}\,V = \frac{r}{a}\,; \qquad \rho = \frac{a(1+\varphi^2)^{\frac{3}{2}}}{2+\varphi^2}\,.$$

$$p = \frac{a\varphi^2}{\sqrt{1+\varphi^2}}\cdot \qquad C = \frac{2a\varphi(1+\varphi^2)}{2+\varphi^2}.$$

If we take a straight line, whose equation is $y = ax$, and lay off the abscissa of any point on the directing circle, and the ordinate of that point on the corresponding radius vector, the point thus determined is a point of the spiral of Archimedes.

In like manner, we may discuss and construct the parabolic spiral, whose equation is $r^2 = 2p\varphi$, and the hyperbolic spiral, whose equation is $r\varphi = m$. The former corresponds to the ordinary parabola, and the latter, to the ordinary hyperbola referred to its asymptotes.

VIII. Transcendental Curves.

Definition.

56. A transcendental curve, is a curve whose equation can only be expressed by the aid of transcendental quantities. The cycloid, and the logarithmic curve are examples of this class of lines.

The Cycloid.

57. The cycloid is a curve that may be generated by a point in the circumference of a circle, when that circle rolls along a straight line. The point is called the *generatrix*, the circle is called the *generating circle*, and the straight line is the *base* of the cycloid. The curve has an infinite number of branches, each corresponding to one revolution of the generating circle.

To find the equation of one branch, APM; let A be the origin, C the centre of the generating circle, and P the place of the generatrix after the circle has rolled through the angle KCP, denoted by φ; denote the co-ordinates, AL and LP, by x and y, and let the radius of the generating circle be represented by r.

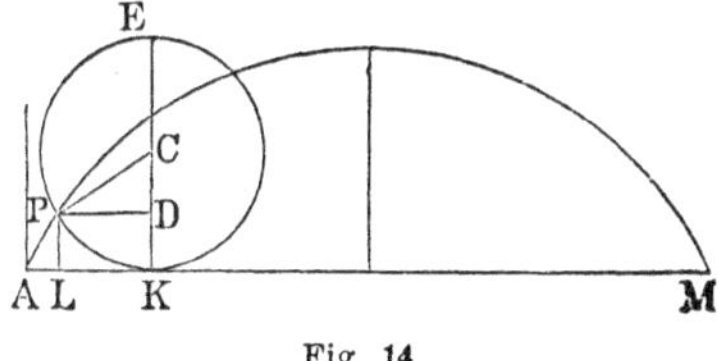

Fig. 14.

From the figure, we have,

$$AL = AK - LK.$$

But, AK is equal to the arc KP, and KP is the arc whose versed sine is y, to the radius r, or r times the arc whose versed sine is $\frac{y}{r}$, to the radius 1; LK is equal to PD, which, from a property of the circle, is equal to $\pm\sqrt{(2r-y)y}$, the upper sign corresponding to the case in which P is to the left of KE, and the lower sign to the case in which P is to the right of KE; and AL is equal to x. Substituting these values, and reducing, we have,

$$x = r\text{versin}^{-1}\frac{y}{r} \mp \sqrt{2ry - y^2};$$

which is the equation sought. From it we see that y can never be less than 0, nor greater than $2r$; we see, also, that y is equal to 0, for $x = 0$, and for $x = 2\pi r$, and that for every value of y, there are two values of x, one exceeding πr as much as the other falls short of it.

Differentiating and reducing, we have,

$$dx = \pm \frac{ydy}{\sqrt{2ry - y^2}},$$

which is the differential equation of the cycloid. From it we deduce the equation,

$$\frac{dy}{dx} = \pm\sqrt{\frac{2r}{y} - 1}, \text{ whence, } \frac{d^2y}{dx^2} = -\frac{r}{y^2}.$$

From the first, we see that the tangent to the curve is vertical for $x = 0$, and $x = 2\pi r$, and that it is horizontal for $x = \pi r$; from the second, we see that the curve is always concave downward.

Substituting, in formulas (6) and (7), Art. 32, we have,

$$S.T = \pm \frac{y''^2}{\sqrt{2ry'' - y''^2}}, \text{ and } S.N = \mp\sqrt{2ry'' - y''^2}.$$

The Logarithmic Curve.

58. The logarithmic curve, is a curve in which any ordinate is equal to the logarithm of the corresponding abscissa. Its equation is, therefore,

$$y = \log x, \text{ or } y = M \,.\, lx;$$

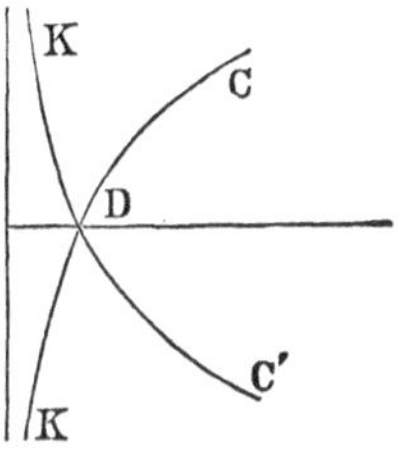

in which M is the modulus of the system.

Fig. 15.

If $M > 0$, y will be negative when $x < 1$, positive when $x > 1$, 0 when $x = 1$, and infinite when $x = 0$.

If $M < 0$, y will be positive when $x < 1$, negative when $x > 1$, 0 when $x = 1$, and infinite when $x = 0$.

In no case can a negative value of x correspond to a point of the curve.

Differentiating, we find,

$$\frac{dy}{dx} = \frac{M}{x}, \text{ and } \frac{d^2y}{dx^2} = -\frac{M}{x^2}.$$

The first expression has the same sign as M, and varies inversely as x. Hence, when $M > 0$, the slope is positive, when $M < 0$, the slope is negative, and in both cases the curve continually approaches parallelism with the axis of x.

The second expression has a sign contrary to that of M, and varies inversely as the square of x. Hence, when $M > 0$, the curve is concave downward, as KDC, and when $M < 0$, it is concave upward, as KDC'.

PART III.

INTEGRAL CALCULUS.

Object of the Integral Calculus.

59. The object of the integral calculus is to pass from a given differential to a function from which it may have been derived. This function is called the *integral of the differential*, and the operation of finding it is called *integration*. The operation of integration is indicated by this sign, $\int$, called the *integral sign*. Integration and differentiation are inverse operations, and their signs, when placed before a quantity, neutralize each other; thus, $\int dx = x$.

A constant quantity connected with a function by the sign of addition, or subtraction, disappears by differentiation, hence, we must add a constant to the integral obtained; and inasmuch as this constant may have any value, it is said to be *arbitrary*. The integral, before the addition of the constant, is called *incomplete*, after its addition it is said to be *complete*. By means of the arbitrary constant, as we shall see hereafter, the integral may be made to satisfy any one reasonable condition.

Nature of an Integral.

60. The differential of a function is the difference between two consecutive states of that function; hence, any state of the function whatever, is equal to some previous, or *initial* state, *plus* the algebraic sum of all the intermediate values of the differential. But this state is the integral, by definition; hence, if the arbitrary constant represents the initial state, the incomplete integral represents the algebraic sum of all the differentials *from* the initial state *up to* any state whatever, and the complete integral represents the initial state, *plus* the algebraic sum of all the differentials from that state up to any state whatever.

Before any value has been assigned to the constant, the integral is said to be *indefinite;* when the value of the constant has been determined, so as to satisfy a particular hypothesis, the integral is said to be *particular;* and when a definite value has been given to the variable, this integral is said to be *definite.*

The values of the variable that correspond to the *initial* and *terminal* states of a definite integral are called *limits,* the former being the *inferior,* and the latter the *superior limit;* the integral is said to lie between these limits. The value of the definite integral may be found from the indefinite integral by substituting for the variable the *inferior* and *superior limits* separately, and then taking the first result from the second. The first result is the integral from any state up to the first limit, and the second is the integral from the same state up to the last limit; hence, the difference is the integral between the limits. The symbol for integrating between the limits is $\int_a^b$, in which a is the inferior, and b the superior limit.

This article will be better understood when we come to its practical application in Articles 78 and following.

Methods of Integration; Simplifications.

61. The methods of integration are not founded on processes of direct reasoning, but are mostly dependent on formulas. The more elementary formulas are deduced by reversing corresponding formulas in the differential calculus; the more complex ones are deduced from these by various transformations and devices, whereby the expressions to be integrated are brought under some known integrable form.

It has been shown (Art. 9), that a constant factor remains unchanged by differentiation; hence, a constant factor may be placed without the sign of integration. It was also shown in the same article, that the differential of the sum of any number of functions is equal to the sum of their differentials; hence, the integral of the sum of any number of differentials is equal to the sum of their integrals. These principles are of continued use in integration.

Fundamental Formulas.

62. If we differentiate the expression $\frac{ax^{n+1}}{n+1}$, we have,

$$d\left(\frac{ax^{n+1}}{n+1}\right) = ax^n dx;$$

reversing the formula, and applying the integral sign to both members,

$$\int ax^n dx = \int d\left(\frac{ax^{n+1}}{n+1}\right).$$

Remembering that the signs of integration and differentiation neutralize each other (Art. 59), and adding a constant to complete the integral, we have,

$$\int ax^n dx = \frac{ax^{n+1}}{n+1} + C \quad . \; . \quad . \; . \; . \; [1]$$

Hence, to integrate a monomial differential,—*drop the differential of the variable, add 1 to the exponent of the variable, divide the result by the new exponent, and add a constant.*

This rule fails when $n = -1$; for, if we apply it, we get a result equal to ∞; in this case the integration may be effected as follows:

From Art. 15, we have,

$$d(alx) = a\frac{dx}{x} = ax^{-1}dx;$$

reversing and proceeding as before, we have,

$$\int ax^{-1}dx = a\int\frac{dx}{x} = alx + C \; . \; . \; . \; . \; . \; [2]$$

From Art. 16, we have,

$$d\left(\frac{a^x}{la}\right) = a^x dx;$$

reversing and proceeding as before, we have,

$$\int a^x dx = \frac{a^x}{la} + C \; . \; . \; . \; . \; . \; . \; . \; . \; . \; . \; [3]$$

In like manner, reversing the formulas in Article 17, lettered from a to h, we have,

$$\int \cos x dx = \sin x + C \; . \; . \; . \; . \; . \; . \; . \; . \; . \; . \; . \; . \; . \; [4]$$

$$\int -\sin x\,dx = \cos x + C \;\ldots\ldots\ldots\ldots \quad [5]$$

$$\int \frac{dx}{\cos^2 x} = \tan x + C \;\ldots\ldots\ldots\ldots \quad [6]$$

$$\int -\frac{dx}{\sin^2 x} = \cot x + C \;\ldots\ldots\ldots\ldots \quad [7]$$

$$\int \sin x\,dx = \text{versin}\, x + C \;\ldots\ldots\ldots\ldots \quad [8]$$

$$\int -\cos x\,dx = \text{coversin}\, x + C \;\ldots\ldots\ldots \quad [9]$$

$$\int \tan x \sec x\,dx = \sec x + C \;\ldots\ldots\ldots\ldots \quad [10]$$

$$\int -\cot x\,\text{cosec}\, x\,dx = \text{cosec}\, x + C \;\ldots\ldots \quad [11]$$

In like manner, from the formulas in Art. 18, lettered fr a' to h', we deduce the following:

$$\int \frac{dy}{\sqrt{1-y^2}} = \sin^{-1} y + C \;\ldots\ldots\ldots \quad [12]$$

$$\int -\frac{dy}{\sqrt{1-y^2}} = \cos^{-1} y + C \;\ldots\ldots\ldots \quad [13]$$

$$\int \frac{dy}{1+y^2} = \tan^{-1} y + C \;\ldots\ldots\ldots\ldots \quad [14]$$

$$\int -\frac{dy}{1+y^2} = \cot^{-1} y + C \;\ldots\ldots\ldots \quad [15]$$

$$\int \frac{dy}{\sqrt{2y - y^2}} = \text{versin}^{-1} y + C \; \ldots\ldots \; [16]$$

$$\int - \frac{dy}{\sqrt{2y - y^2}} = \text{coversin}^{-1} y + C \; \ldots \; [17]$$

$$\int \frac{dy}{y\sqrt{y^2 - 1}} = \sec^{-1} y + C \; \ldots\ldots\ldots \; [18]$$

$$\int - \frac{dy}{y\sqrt{y^2 - 1}} = \text{cosec}^{-1} y + C \; \ldots . \; [19]$$

in formulas (12), (13), (14), (15), (18), (19), we make $\frac{bx}{a}$, whence $dy = \frac{bdx}{a}$, and reduce, we have,

$$\int \frac{dx}{\sqrt{a^2 - b^2x^2}} = \frac{1}{b} \sin^{-1} \frac{bx}{a} + C \; \ldots . \; [20]$$

$$\int - \frac{dx}{\sqrt{a^2 - b^2x^2}} = \frac{1}{b} \cos^{-1} \frac{bx}{a} + C \; \ldots \; [21]$$

$$\int \frac{dx}{a^2 + b^2x^2} = \frac{1}{ab} \tan^{-1} \frac{bx}{a} + C \; \ldots\ldots \; [22]$$

$$\int - \frac{dx}{a^2 + b^2x^2} = \frac{1}{ab} \cot^{-1} \frac{bx}{a} + C \; \ldots \; [23]$$

$$\int \frac{dx}{x\sqrt{b^2x^2 - a^2}} = \frac{1}{a} \sec^{-1} \frac{bx}{a} + C \; \ldots . \; [24]$$

$$\int - \frac{dx}{x\sqrt{b^2x^2 - a^2}} = \frac{1}{a}\,\text{cosec}^{-1}\frac{bx}{a} + C\,. \quad \ldots [25]$$

From (16) and (17), by substituting $\frac{2b^2}{a^2}x$ for y, and reducing, we have,

$$\int \frac{dx}{\sqrt{a^2x - b^2x^2}} = \frac{1}{b}\,\text{versin}^{-1}\frac{2b^2}{a^2}x + C \,.\,.\,.\,.\,. [26]$$

$$\int - \frac{dx}{\sqrt{a^2x - b^2x^2}} = \frac{1}{b}\,\text{coversin}^{-1}\frac{2b^2}{a^2}x + C \,.\,. [27]$$

Formulas (1) to (19) are the elementary forms, to one of which we endeavor to reduce every case of integration; the processes of the integral calculus are little else than a succession of transformations and devices, by which this reduction is affected.

In the following examples, and, as a general thing, throughout the book, the incomplete integral is given, it being understood that an arbitrary constant is to be added in every case.

EXAMPLES.

Formulas (1) to (3).

1. $dy = ax^3dx.$ *Ans.* $y = \frac{ax^4}{4}$,

2. $dy = bx^{\frac{1}{2}}dx.$ *Ans.* $y = \frac{2}{3}bx^{\frac{3}{2}}$,

3. $dy = 3x^{-4}dx.$ *Ans.* $y = -x^{-3}$.

4. $dy = 2x^{\frac{5}{7}}dx.$ *Ans.* $y = \frac{7}{6}x^{\frac{12}{7}}.$

5. $dy = -\frac{2}{3}x^{-\frac{6}{5}}dx.$ *Ans.* $y = \frac{10}{3}x^{-\frac{1}{5}}.$

6. $dy = \left(\frac{5}{2}ax^{\frac{3}{2}} - \frac{3}{2}bx^{\frac{1}{2}}\right)dx.$ *Ans.* $y = ax^{\frac{5}{2}} - bx^{\frac{3}{2}}.$

7. $dy = \left(\frac{12}{x^2} - \frac{3}{x^4}\right)dx.$ *Ans.* $y = -\frac{12}{x} + \frac{1}{x^3}.$

8. $dy = \frac{1}{2}x^{-\frac{1}{2}}dx.$ *Ans.* $y = x^{\frac{1}{2}}.$

9. $dy = 3x^{-1}dx.$ *Ans.* $y = 3lx.$

10. $dy = (2x^{\frac{2}{3}} + x^{-1})dx.$ *Ans.* $y = \frac{6}{5}x^{\frac{5}{3}} + lx.$

11. $dy = e^{\sin x}\cos x dx.$ *Ans.* $y = e^{\sin x}.$

12. $dy = -e^{\cos x}\sin x dx.$ *Ans.* $y = e^{\cos x}.$

Formulas (4) to (11).

13. $dy = 2\cos 2x dx.$ *Ans.* $y = \sin 2x.$

14. $dy = -\frac{1}{2}\sin(x^2) \times x dx.$ *Ans.* $y = \frac{1}{4}\cos(x^2)$

15. $dy = \frac{dx}{\cos^2(\frac{1}{2}x)}.$ *Ans.* $y = 2\tan\left(\frac{1}{2}x\right)$

16. $dy = -\frac{x dx}{\sin^2(3x^2)}.$ *Ans.* $y = \frac{1}{6}\cot(3x^2)$

17. $dy = \sin(ax)dx.$ *Ans.* $y = \frac{1}{a}\text{versin}(ax).$

18. $dy = -\cos\left(\frac{1}{2}x^2\right) \times xdx.$ *Ans.* $y = \text{coversin}\left(\frac{1}{2}x^2\right).$

Formulas (12) to (19).

19. $dy = \frac{2dx}{\sqrt{1-4x^2}}.$ *Ans.* $y = \sin^{-1}(2x).$

20. $dy = \frac{2xdx}{\sqrt{1-x^4}}.$ *Ans.* $y = \sin^{-1}(x^2).$

21. $dy = \frac{x^{\frac{1}{2}}dx}{\sqrt{2-4x^3}} = \frac{1}{3}\frac{d(x\sqrt{2x})}{\sqrt{1-2x^3}}.$

Ans. $y = \frac{1}{3}\sin^{-1}(x\sqrt{2x}).$

22. $dy = -\frac{dx}{\sqrt{x-4x^2}} = -\frac{d(2x^{\frac{1}{2}})}{\sqrt{1-4x}}$

Ans. $y = \cos^{-1}(2\sqrt{x}).$

23. $dy = \frac{xdx}{1+x^4} = \frac{1}{2}\frac{d(x^2)}{1+x^4}.$ *Ans.* $y = \frac{1}{2}\tan^{-1}x^2.$

24. $dy = -\frac{2dx}{1+x^2}.$ *Ans.* $y = 2\cot^{-1}x.$

25. $dy = \frac{dx}{\sqrt{6x-9x^2}}.$ *Ans.* $y = \frac{1}{3}\text{versin}^{-1}3x.$

Formulas (20) to (25).

26. $dy = \frac{dx}{\sqrt{2 - 9x^2}}$; $(a = \sqrt{2},\ b = 3)$.

Ans. $y = \frac{1}{3}\sin^{-1}\frac{3x}{\sqrt{2}}$.

27. $dy = \frac{-\frac{2}{3}xdx}{\sqrt{2 - 5x^4}}$; $(a = \sqrt{2},\ b = \sqrt{5},\ u = x^2)$.

Ans. $y = \frac{1}{3\sqrt{5}}\cos^{-1}\frac{x^2\sqrt{5}}{\sqrt{2}}$.

28. $dy = \frac{3dx}{4 + 9x^2}$; $(a = 2,\ b = 3)$.

Ans. $y = \frac{1}{2}\tan^{-1}\frac{3x}{2}$.

29. $dy = -\frac{2dx}{4 + x^2}$; $(a = 2,\ b = 1)$.

Ans. $y = \cot^{-1}\frac{x}{2}$.

30. $dy = \frac{2dx}{x\sqrt{3x^2 - 5}}$; $(a = \sqrt{5},\ b = \sqrt{3})$.

Ans. $y = \frac{2}{\sqrt{5}}\sec^{-1}\frac{x\sqrt{3}}{\sqrt{5}}$.

31. $dy = -\frac{3dx}{\sqrt{5x^4 - 2x^2}}$; $(a = \sqrt{2},\ b = \sqrt{5})$.

Ans. $y = \frac{3}{\sqrt{2}}\operatorname{cosec}^{-1}\frac{x\sqrt{5}}{\sqrt{2}}$.

32. $dy = \frac{dx}{1 + 5x^2}$. *Ans.* $y = \frac{1}{\sqrt{5}}\tan^{-1}x\sqrt{5}$

33. $dy = \frac{-2x^{-1}dx}{\sqrt{14x^2 - 3}}$. *Ans.* $y = \frac{2}{\sqrt{3}}\text{cosec}^{-1}\frac{x\sqrt{14}}{\sqrt{3}}$.

Formulas (26) and (27).

34. $dy = \frac{2xdx}{\sqrt{4x^2 - 9x^4}}$; $(a = 2,\ b = 3)$.

Ans. $y = \frac{1}{3}\text{versin}^{-1}\frac{9x^2}{2}$.

35. $dy = \frac{-3axdx}{\sqrt{x^2 - 2x^4}}$; $(a = 1,\ b = \sqrt{2})$.

Ans. $y = \frac{3a}{2\sqrt{2}}\text{coversin}^{-1}4x^2$.

EXAMPLES IN SUCCESSIVE INTEGRATION.

36. $d^3y = ax\,dx^3$.

Dividing by dx^2,

$$\frac{d^3y}{dx^2} = axdx, \text{ or } d\left(\frac{d^2y}{dx^2}\right) = axdx.$$

Integrating, we have,

$$\frac{d^2y}{dx^2} = \frac{ax^2}{2} + C.$$

Multiplying by dx, and integrating again, we find,

$$\frac{dy}{dx} = \frac{ax^3}{6} + Cx + C'.$$

Multiplying again by dx, and integrating,

$$y = \frac{ax^4}{24} + \frac{Cx^2}{2} + C'x + C''.$$

37. $d^4y = -x^{-4}dx^4$.

As before, we have,

$$\frac{d^3y}{dx^3} = \frac{x^{-3}}{3} + C;$$

$$\frac{d^2y}{dx^2} = -\frac{x^{-2}}{6} + Cx + C';$$

$$\frac{dy}{dx} = \frac{x^{-1}}{6} + \frac{Cx^2}{2} + C'x + C'';$$

and finally,

$$y = \frac{1}{6}lx + \frac{Cx^3}{6} + \frac{C'x^2}{2} + C''x + C'''.$$

38. $d^3y = mdx^3$.

$$Ans.\ y = \frac{mx^3}{6} + \frac{Cx^2}{2} + C'x + C''$$

39. $d^3y = x^{\frac{1}{2}}dx^3$.

$$Ans.\ y = \frac{8}{105}x^{\frac{7}{2}} + \frac{Cx^2}{2} + C'x + C''$$

40. $d^2y = sx^2dx^2$.

$$Ans.\ y = \frac{sx^4}{12} + Cx + C'.$$

Integration by Parts.

63. If u and v are functions of x, we have, from Art. 10

$$d(uv) = udv + vdu.$$

Integrating and transposing, we have,

$$\int udv = uv - \int vdu \quad . \; . \; . \; . \; . \quad [28]$$

This is called the *formula for integration by parts;* it enables us to integrate an expression of the form udv whenever we can integrate an expression of the form vdu. It is much used in reducing integrals to known forms.

EXAMPLES.

1. $dy = xlx\,dx;$

Let $lx = u$, and $xdx = dv;$

$$\therefore du = \frac{dx}{x}, \text{ and } v = \frac{x^2}{2}.$$

Substituting in the formula, we have,

$$y = \frac{x^2 lx}{2} - \int \frac{xdx}{2} = \frac{x^2 lx}{2} - \frac{x^2}{4}.$$

2. $dy = \dfrac{dx\sqrt{1-x^2}}{x^2}$

Let $\sqrt{1-x^2} = u$, and $\dfrac{dx}{x^2} = dv;$

$$\therefore du = -\frac{xdx}{\sqrt{1-x^2}}, \text{ and } v = -\frac{1}{x};$$

hence, we have, from the formula,

$$y = -\frac{\sqrt{1-x^2}}{x} - \int \frac{dx}{\sqrt{1-x^2}} = -\frac{\sqrt{1-x^2}}{x} - \sin^{-1}x.$$

Additional Formulas.

64. 1°. Formula (1) may be extended to cover the case of a binomial differential of the form,

$$dy = (a + bx^n)^p x^{n-1} dx,$$

in which the exponent of the variable without the parenthesis is 1 less than the exponent of the variable within.

$$\text{Let } a + bx^n = z; \quad \therefore\ bnx^{n-1}dx = dz, \text{ or } x^{n-1}dx = \frac{dz}{bn}.$$

Substituting, and integrating, we have,

$$y = \int (a + bx^n)^p x^{n-1} dx = \int \frac{z^p dz}{bn} = \frac{z^{p+1}}{bn(p+1)}.$$

Replacing z by its equal, $(a + bx^n)$, we have,

$$\int (a + bx^n)^p x^{n-1} dx = \frac{(a + bx^n)^{p+1}}{bn(p+1)} + C \quad \ldots\ldots \quad [29]$$

Hence, to integrate a binomial differential of the proposed form, *add* 1 *to the exponent of the parenthesis, divide the result by the new exponent into the product of the coefficient and the exponent of the variable within the parenthesis.*

It is to be observed that a constant factor may be set aside during the process of integration, and then introduced, as explained in Article 61.

EXAMPLES.

1. $dy = (1 + x^2)^{\frac{1}{2}}xdx.$ *Ans.* $y = \frac{(1 + x^2)^{\frac{3}{2}}}{3}$

2. $dy = \frac{xdx}{\sqrt{a^2 + x^2}}.$ *Ans.* $y = \sqrt{a^2 + x^2}$

3. $dy = 5x^3\,dx\sqrt{7 + 3x^4}.$ *Ans.* $y = \frac{5}{18}(7 + 3x^4)^{\frac{3}{2}}.$

4. $dy = (2 + 3x^2)^3 xdx.$ *Ans.* $y = \frac{1}{24}(2 + 3x^2)^4.$

5. $dy = \frac{2xdx}{(x^2 + 1)^2}.$ *Ans.* $y = -\frac{1}{(x^2 + 1)}.$

2°. The preceding formula fails when $p = -1$; in that case, we have,

$$\int (a + bx^n)^{-1}x^{n-1}dx = \int \frac{z^{-1}dz}{bn} = \frac{1}{bn}\int \frac{dz}{z} = \frac{1}{bn}lz + C.$$

Replacing z by its value,

$$\int (a + bx^n)^{-1}x^{n-1}dx = \frac{1}{bn}l(a + bx^n) + C \; . \; . \; . \; . \; . \; [30]$$

EXAMPLES.

1. $dy = 3(3 + 4x^3)^{-1}x^2dx.$ *Ans.* $y = \frac{1}{4}l(3 + 4x^3).$

2. $dy = \frac{dx}{x + a}.$ *Ans.* $y = l(x + a)$

3. $dy = \frac{2dx}{x + 4}.$ *Ans.* $y = 2\,l(x + 4)$

When the differential expression is a fraction in which the numerator is equal to the differential of the denominator multiplied by a constant, its integral is equal to that constant into the Napierian logarithm of the denominator.

4. $dy = \frac{x^2 dx}{a^3 - x^3}$. *Ans.* $y = -\frac{1}{3} l(a^3 - x^3)$.

5. $dy = \frac{3x^2 + 2x + 1}{x^3 + x^2 + x + 1} dx$.

Ans. $y = l(x^3 + x^2 + x + 1)$

3°. Let it be required to integrate the expression,

$$dy = \frac{dx}{\sqrt{x^2 \pm a^2}};$$

assume,

$$x^2 \pm a^2 = z^2; \text{ whence, } xdx = zdz;$$

adding zdx to both members and factoring, we have,

$$(x + z)dx = z(dx + dz);$$

hence,

$$\frac{dx}{z} = \frac{dx}{\sqrt{x^2 \pm a^2}} = \frac{dx + dz}{x + z} = \frac{d(x + z)}{x + z};$$

applying the principle just deduced, we have,

$$\int \frac{dx}{\sqrt{x^2 \pm a^2}} = l(x + z) + C.$$

Replacing z by its value, we have, finally,

$$\int \frac{dx}{\sqrt{x^2 \pm a^2}} = l(x + \sqrt{x^2 \pm a^2}) + C \; . \; . \; . \; . \; . \; [31]$$

EXAMPLES.

1. $dy = \frac{dx}{\sqrt{4x^2 - 7}} = \frac{1}{2}\frac{dx}{\sqrt{x^2 - \frac{7}{4}}}$.

$$Ans.\ y = \frac{1}{2} l(x + \sqrt{x^2 - \tfrac{7}{4}})$$

2. $dy = \frac{3dx}{\sqrt{x^2 - 5}}$. $\qquad Ans.\ y = 3l(x + \sqrt{x^2 - 5})$.

4°. Let it be required to integrate the expression,

$$dy = \frac{dx}{\sqrt{2ax + x^2}};$$

since, $dx = d(a + x)$, and $2ax + x^2 = (a + x)^2 - a^2$, we have,

$$\int \frac{dx}{\sqrt{2ax + x^2}} = \int \frac{d(a + x)}{\sqrt{(a + x)^2 - a^2}},$$

which can be integrated by Formula (31); hence,

$$\int \frac{dx}{\sqrt{2ax + x^2}} = l\left\{a + x + \sqrt{2ax + x^2}\right\} + C \quad . \quad . \quad [32]$$

EXAMPLES.

1. $dy = \frac{2dx}{\sqrt{3x + 4x^2}} = \frac{dx}{\sqrt{\frac{3}{4}x + x^2}}$.

$$Ans.\ y = l\left(x + \frac{3}{8} + \sqrt{\tfrac{3}{4}x + x^2}\right)$$

2. $dy = \dfrac{3dx}{\sqrt{5x + 9x^2}}$. *Ans.* $y = l\left(x + \dfrac{5}{18} + \sqrt{\tfrac{5}{9}x + x^2}\right)$.

5°. Let it be required to integrate the expression,

$$dy = \frac{dx}{a^2 - x^2}.$$

Factoring, we have,

$$\frac{dx}{a^2 - x^2} = \frac{1}{2a}\left\{\frac{dx}{a + x} + \frac{dx}{a - x}\right\}.$$

Integrating, we have,

$$\int \frac{dx}{a^2 - x^2} = \frac{1}{2a}\Big(l(a + x) - l(a - x)\Big).$$

But the difference of the logarithms of two quantities is equal to the logarithm of their quotient; hence,

$$\int \frac{dx}{a^2 - x^2} = \frac{1}{2a}\, l\, \frac{a + x}{a - x} + C \; . \; . \; . \; . \; . \; [33]$$

and in like manner,

$$\int \frac{dx}{x^2 - a^2} = \frac{1}{2a}\, l\, \frac{x - a}{x + a} + C \; . \; . \; . \; . \; . \; [34]$$

EXAMPLES.

1. $dy = \dfrac{dx}{9 - x^2}$. *Ans.* $y = \dfrac{1}{6}\, l\, \dfrac{3 + x}{3 - x}$

2. $dy = \dfrac{dx}{x^2 - 4}$ *Ans.* $y = \dfrac{1}{4}\, l\, \dfrac{x - 2}{x + 2}$

3. Given $\frac{d^2y}{dx^2} = y$, to find the relation between y and x.

Multiplying both members by $2dy$,

$$\frac{2dy\,d^2y}{dx^2} = 2ydy, \text{ or } \frac{d(dy^2)}{dx^2} = 2ydy.$$

Integrating,

$$\frac{dy^2}{dx^2} = y^2 + C; \quad \therefore\ dx = \frac{dy}{\sqrt{y^2 + C}}.$$

Integrating by Formula 31,

$$x = l(y + \sqrt{y^2 + C}) + C'.$$

4. Given $\frac{d^2s}{dt^2} = -n^2s$, to find the relation between s and t, t being the independent variable.

Multiplying by $2ds$,

$$\frac{2dsd^2s}{dt^2} = -2n^2sds.$$

Integrating,

$$\frac{ds^2}{dt^2} = C - n^2s^2; \quad \therefore\ dt = \frac{ds}{\sqrt{C - n^2s^2}}.$$

Integrating by Formula (20),

$$t = \frac{1}{n}\sin^{-1}\frac{ns}{\sqrt{C}} + C'.$$

Rational and Entire Differentials.

65. An algebraic function is *rational and entire* when it contains no radical or fractional part that involves the variable. If the indicated operations be performed, every rational and entire differential will be reduced to the form of a monomial, or polynomial differential, each term of which can be integrated by formulas (1) or (2). Thus, if the indicated operations be performed, in the expression,

$$dy = \left(x - \frac{1}{2}\right)\left(x^2 + \frac{3}{4}\right)x^3 dx,$$

we have,

$$dy = \frac{1}{8}\left(8x^6 - 4x^5 + 6x^4 - 3x^3\right)dx.$$

Hence, by integration,

$$y = \frac{1}{8}\left(\frac{8}{7}x^7 - \frac{2}{3}x^6 + \frac{6}{5}x^5 - \frac{3}{4}x^4\right)$$

$$= \left(\frac{1}{7}x^3 - \frac{1}{12}x^2 + \frac{3}{20}x - \frac{3}{32}\right)x^4.$$

In like manner, any rational and entire differential may be integrated.

Rational Fractions.

66. A rational fraction, is a fraction whose terms are rational and entire. When a rational fraction is the differential of a function, and its numerator is of a higher degree than its denominator, it may, by the process of division, be separated into two differentials, one of which is rational and entire and the other a rational fraction, in which the numerator is of a lower degree than the denominator. The former may be integrated as in the last article; it remains to be shown that the latter can be integrated whenever the denominator can be resolved into binomial factors of the first degree with respect to the variable. The method of integration consists in resolving the differential coefficients into partial fractions, by some of the known methods, then multiplying each by the differential of the variable and integrating the polynomial result. There are four cases, depending on the method of resolving the differential coefficient into partial fractions. Each case will be illustrated by examples.

First Case.—*When the binomial factors of the denominator are unequal.*

Let us have,

$$dy = \frac{(2x-5)dx}{x^3 - 6x^2 + 11x - 6} = \frac{2x-5}{(x-1)(x-2)(x-3)}dx.$$

Assume the identical equation,

$$\frac{2x-5}{(x-1)(x-2)(x-3)} = \frac{A}{x-1} + \frac{B}{x-2} + \frac{C}{x-3} \quad \ldots\ldots (1)$$

Clearing of fractions.

$$2x - 5 = A(x-2)(x-3) + B(x-1)(x-3) + C(x-1)(x-2) \quad \ldots\ldots (2)$$

In order to find A, B, and C, we might perform the operations indicated in (2), equate the coefficients of the like powers of x, and solve the resulting equations; but there is a simpler method in cases like this, depending on the fact that (2) is true for all values of x.

Making $x = 1$, we find, $-3 = 2A$; $\therefore A = -\frac{3}{2}$.

Making $x = 2$, we find, $-1 = -B$; $\therefore B = 1$.

Making $x = 3$, we find, $1 = 2C$; $\therefore C = \frac{1}{2}$.

Substituting these in (1), multiplying by dx, and integrating by Formula 30, we have,

$$\int\frac{(2x-5)dx}{(x-1)(x-2)(x-3)} = -\frac{3}{2}\int\frac{dx}{x-1} + \int\frac{dx}{x-2}$$

$$+\frac{1}{2}\int\frac{dx}{x-3} = -\frac{3}{2}l(x-1) + l(x-2) + \frac{1}{2}l(x-3).$$

Reducing the result to its simplest form, in accordance with the elementary principles of logarithms, we have, finally,

$$y = l\frac{(x-2)(x-3)^{\frac{1}{2}}}{(x-1)^{\frac{3}{2}}}.$$

EXAMPLES.

1. $dy = \dfrac{(5x+1)dx}{x^2+x-2} = \dfrac{(5x+1)dx}{(x-1)(x+2)}$.

Ans. $y = 2l(x-1) + 3l(x+2) = l\left[(x-1)^2 \times (x+2)^3\right]$.

2. $dy = \dfrac{(x-1)dx}{x^2+6x+8} = \dfrac{(x-1)dx}{(x+2)(x+4)}$.

Ans. $y = l\dfrac{(x+4)^{\frac{5}{2}}}{(x+2)^{\frac{3}{2}}}$.

3. $dy = \dfrac{(2x+3)dx}{x^3+x^2-2x} = \dfrac{(2x+3)dx}{x(x-1)(x+2)}$.

Ans. $y = l\dfrac{(x-1)^{\frac{5}{3}}}{x^{\frac{3}{2}}(x+2)^{\frac{1}{6}}}$.

4. $dy = \dfrac{(3x^2-1)dx}{x(x-1)(x+1)}$.

Ans. $y = l(x^3-x)$.

Second Case.—When some of the binomial factors are equal.

In this case there are as many partial fractions as there are binomial factors in the denominator; but the denominators of those corresponding to factors of the form $(x-a)^m$, are respectively of the form $(x-a)^m$, $(x-a)^{m-1}$, $(x-a)^{m-2}$, etc., down to $x-a$.

Let us assume,

$$dy = \frac{(x^2+x)dx}{(x-2)^2(x-1)};$$

assuming the identical equation,

$$\frac{x^2+x}{(x-2)^2(x-1)} = \frac{A}{(x-2)^2} + \frac{B}{x-2} + \frac{C}{x-1} \quad \ldots\ldots (1)$$

and clearing of fractions, we have,

$$x^2 + x = A(x-1) + B(x-2)(x-1) + C(x-2)^2 \ldots (2)$$

Here again we might find the values of A, B, and C, by the ordinary method of indeterminate coefficients; but it will be simpler to proceed as follows:

Making $x = 2$, we find, $A = 6$;

Making $x = 1$, we find, $C = 2$;

giving to A and C their values in (2), and then making $x = 0$, we find, $0 = -6 + 2B + 8 \quad \therefore B = -1$.

Substituting in (1) and multiplying by dx, we have,

$$\frac{(x^2+x)dx}{(x-2)^2(x-1)} = \frac{6dx}{(x-2)^2} - \frac{dx}{x-2} + \frac{2dx}{x-1}.$$

The first partial fraction can be integrated by Formula 29, and the others by Formula 30; hence,

$$y = -\frac{6}{x-2} - l(x-2) + 2l(x-1) = -\frac{6}{x-2} + l\frac{(x-1)^2}{x-2}.$$

EXAMPLES.

1. $dy = \dfrac{(x^2-2)dx}{x^3(x-1)}.$

In this example the equal binomial factors are of the form $(x-0)$ or x, and the assumed identical equation is,

$$\frac{x^2-2}{x^3(x-1)} = \frac{A}{x^3} + \frac{B}{x^2} + \frac{C}{x} + \frac{D}{x-1} \quad \ldots\ldots \quad (1)$$

Clearing of fractions, and performing indicated operations, we have,

$$x^2 - 2 = Ax - A + Bx^2 - Bx + Cx^3 - Cx^2 + Dx^3.$$

Equating the coefficients of like powers in the two members, and solving the resulting group, we have, $A = 2$, $B = 2$, $C = 1$, and $D = -1$; substituting in (1), and proceeding as before, we have,

$$y = -\frac{2x+1}{x^2} + l\frac{x}{x-1}.$$

2. $dy = \dfrac{x^2dx}{(x+2)^2(x+1)}.$ *Ans.* $y = \dfrac{4}{x+2} + l(x+1)$

3. $dy = \dfrac{xdx}{(x+2)(x+3)^2}.$

Ans. $y = -\dfrac{3}{x+3} + l\left(\dfrac{x+3}{x+2}\right)^2$

Third Case.—When some of the factors are imaginary but unequal.

In this case the product of each pair of imaginary factors is of the form $(x - a)^2 + b^2$; instead of assuming a partial fraction for each imaginary factor, we assume for each pair of such factors a fraction of the form,

$$\frac{A + Bx}{(x - a)^2 + b^2}.$$

Let $dy = \dfrac{xdx}{(x + 1)(x^2 + 1)}$.

Assume the identical equation,

$$\frac{x}{(x + 1)(x^2 + 1)} = \frac{A + Bx}{x^2 + 1} + \frac{C}{x + 1} \quad \ldots\ldots \quad (1)$$

Clearing of fractions, and reducing,

$$x = Bx^2 + (A + B)x + A + Cx^2 + C \quad \ldots\ldots \quad (2)$$

Equating the coefficients of the like powers of x in the two members, and solving the resulting equations, we have,

$$A = \frac{1}{2}, \quad B = \frac{1}{2}, \text{ and } C = -\frac{1}{2}.$$

Substituting in (1), and multiplying by dx, we have,

$$\frac{xdx}{(x + 1)(x^2 + 1)} = \frac{1}{2}\frac{(1 + x)dx}{x^2 + 1} - \frac{1}{2}\frac{dx}{x + 1}$$

$$= \frac{1}{2}\left(\frac{dx}{x^2 + 1} + \frac{xdx}{x^2 + 1} - \frac{dx}{x + 1}\right).$$

Integrating by Formulas (14) and (30), and reducing, we have,

$$y = \frac{1}{2}\tan^{-1}x + \frac{1}{2}l\frac{(x^2 + 1)^{\frac{1}{2}}}{x + 1}.$$

Fourth Case.—When some of the binomial factors are imaginary and equal.

In this case, the denominator will contain one or more factors of the form $[(x - a)^2 + b^2]^n$; in determining the partial fractions, we combine the methods used in the *second* and *third* cases.

Let us assume,

$$dy = \frac{x^3 dx}{(x^2 - 2x + 2)^2}.$$

Assume the identical equation,

$$\frac{x^3}{(x^2 - 2x + 2)^2} = \frac{Ax + B}{(x^2 - 2x + 2)^2} + \frac{Cx + D}{x^2 - 2x + 2} \ldots\ldots (1)$$

Clearing of fractions, and proceeding as before, we find,

$$A = 2, \ B = -4, \ C = 1, \text{ and } D = 2.$$

Substituting these in (1), multiplying by dx, and separating the fractions, we have,

$$dy = \frac{2xdx}{(x^2 - 2x + 2)^2} - \frac{4dx}{(x^2 - 2x + 2)^2} + \frac{xdx}{x^2 - 2x + 2}$$

$$+ \frac{2dx}{x^2 - 2x + 2} \ldots\ldots (2)$$

Making $(x - 1)^2 + 1 = z^2 + 1$, whence $x - 1 = z$, and $dx = dz$, we have,

$$dy = \frac{2(z + 1)dz}{(z^2 + 1)^2} - \frac{4dz}{(z^2 + 1)^2} + \frac{(z + 1)dz}{z^2 + 1} + \frac{2dz}{z^2 + 1},$$

or,

$$dy = \frac{2zdz}{(z^2 + 1)^2} - \frac{2dz}{(z^2 + 1)^2} + \frac{zdz}{z^2 + 1} + \frac{3dz}{z^2 + 1}.$$

The *first*, *third*, and *fourth* terms can be integrated by known formulas; the second is a particular case of the form $\frac{dz}{(z^2 + 1)^n}$, which can be integrated by the aid of Formula D, yet to be deduced.

Integrating the *first*, *third*, and *fourth*, and indicating the integration of the *second*, we have,

$$y = -\frac{1}{(z^2 + 1)} - 2\int \frac{dz}{(z^2 + 1)^2} + \frac{1}{2}l(z^2 + 1) + 3\tan^{-1}z.$$

From what precedes we infer that all rational differentials are integrable; consequently, all differentials that can be made rational in terms of a new variable are also integrable.

Integration by Substitution, and Rationalization.

67. An irrational differential may sometimes be made rational, by substituting for the variable some function of an auxiliary variable; when this can be done, the integration may be effected by the methods of Articles 65 and 66. When the differential cannot be rationalized in terms of an auxiliary variable, it may sometimes be reduced to one of the elementary forms, and then integrated. The method of proceeding is best illustrated by examples.

When the only Irrational Parts are Monomial.

68. When the only irrational parts are monomial, a differential can be made rational by substituting for the primitive variable a new variable raised to a power whose exponent is the least common multiple of all the indices in the expression.

Let $$dy = \frac{(x^{\frac{1}{2}} - x^{\frac{2}{3}})dx}{1 + x^{\frac{1}{3}}} \quad \ldots\ldots (1)$$

The least common multiple of the indices is 6; making $x = z^6$, we have, $x^{\frac{1}{2}} = z^3$, $x^{\frac{2}{3}} = z^4$, $x^{\frac{1}{3}} = z^2$, and $dx = 6z^5dz$: these substituted in (1), give,

$$dy = \frac{(z^3 - z^4)}{1 + z^2} 6z^5 dz = -6\frac{z^9 - z^8}{z^2 + 1}dz.$$

Performing the division indicated, we have,

$$dy = -6\left(z^7 - z^6 - z^5 + z^4 + z^3 - z^2 - z + 1 + \frac{z}{z^2+1} - \frac{1}{z^2+1}\right)dz.$$

Integrating by known methods, and replacing z by its value, $x^{\frac{1}{6}}$, we have, finally,

$$y = -\frac{3}{4}x^{\frac{4}{3}} + \frac{6}{7}x^{\frac{7}{6}} + x - \frac{6}{5}x^{\frac{5}{6}} - \frac{3}{2}x^{\frac{2}{3}} + 2x^{\frac{1}{2}} + 3x^{\frac{1}{3}} - 6x^{\frac{1}{6}} - 3l(1 + x^{\frac{1}{3}}) + 6\tan^{-1}x^{\frac{1}{6}}.$$

When the only irrational parts are fractional powers of a binomial of the first degree, the differential can be rationalized by the same rule, and consequently can be integrated.

Let us take the expression,

$$dy = \left(x + \sqrt{x+2} + \sqrt[3]{x+2}\right)dx \quad \ldots\ldots (2)$$

Assume,

$$x + 2 = z^6, \text{ whence, } x = z^6 - 2, \text{ and } dx = 6z^5dz.$$

Substituting in (2), we have,

$$dy = (z^6 - 2 + z^3 + z^2)6z^5dz = 6(z^{11} - 2z^5 + z^8 + z^7)dz.$$

Integrating, and substituting for z its value, $(x + 2)^{\frac{1}{6}}$, we have,

$$y = \frac{1}{2}(x + 2)^2 - 2(x + 2) + \frac{2}{3}(x + 2)^{\frac{3}{2}} + \frac{3}{4}(x + 2)^{\frac{4}{3}}.$$

Binomial Differentials.

69. Every binomial differential can be reduced to the form,

$$dy = A(a + bx^n)^{\frac{r}{s}}x^{m-1}dx \quad . \; . \; . \; . \; . \quad (1)$$

in which m, n, r, and s are whole numbers, and n positive.

Thus, the expression, $(x^{-\frac{1}{2}} + 2x^{-\frac{1}{4}})^{-\frac{2}{3}}x^{\frac{1}{6}}dx$, can be reduced to the form $(1 + 2x^{\frac{1}{4}})^{-\frac{2}{3}}x^{\frac{1}{2}}dx$ by simply removing the factor $x^{-\frac{1}{2}}$ from under the radical sign. If, in this result, we make $x = z^4$, whence, $dx = 4z^3dz$, it will become, $4(1 + 2z)^{-\frac{2}{3}}z^5dz$, which is of the required form. In like manner, any similar expression may be reduced to that form. The constant factor, A, may be omitted during integration, and the exponent of the parenthesis, $\frac{r}{s}$, may be represented by a single letter p, as is done in Article 70. After integration, the factor may be replaced, and the value of $\frac{r}{s}$ may be substituted for p.

FIRST CRITERION.

Form (1) may be integrated by some of the methods explained in Articles 65 and 66, when $\frac{r}{s}$ is a whole number.

SECOND CRITERION.

Form (1) can be integrated when $\frac{m}{n}$ is a whole number.

For let us assume $(a + bx^n) = z^s$, or, $z = (a + bx^n)^{\frac{1}{s}}$; we have, by solution and differentiation,

$$x = \left(\frac{z^s - a}{b}\right)^{\frac{1}{n}};\ x^m = \left(\frac{z^s - a}{b}\right)^{\frac{m}{n}};$$

and,

$$x^{m-1}dx = \frac{s}{bn}\left(\frac{z^s - a}{b}\right)^{\frac{m}{n}-1} z^{s-1}dz.$$

Substituting in (1), we find,

$$\left(a + bx^n\right)^{\frac{r}{s}} x^{m-1}dx = \frac{s}{bn}\left(\frac{z^s - a}{b}\right)^{\frac{m}{n}-1} z^{r+s-1}dz \quad . . . (2)$$

Which expression is integrable, as was shown above, when $\frac{m}{n}$ is a whole number.

THIRD CRITERION.

Form (1) can be integrated when $\frac{m}{n} + \frac{r}{s}$, is a whole number.

For let us assume,

$$a + bx^n = z^s x^n, \text{ or, } z = \left(\frac{a + bx^n}{x^n}\right)^{\frac{1}{s}}.$$

Solving and differentiating, we have,

$$x = \left(\frac{a}{z^s - b}\right)^{\frac{1}{n}} = a^{\frac{1}{n}}(z^s - b)^{-\frac{1}{n}};\ x^m = a^{\frac{m}{n}}(z^s - b)^{-\frac{m}{n}};$$

and,

$$x^{m-1}dx = -\frac{s}{n}a^{\frac{m}{n}}(z^s - b)^{-\frac{m}{n}-1}z^{s-1}dz.$$

Substituting in (1), and reducing, we find,

$$(a + bx^n)^{\frac{r}{s}}x^{m-1}dx = -\frac{s}{n}a^{\frac{m}{n}+\frac{r}{s}}(z^s - b)^{-\frac{m}{n}-\frac{r}{s}-1}z^{r+s-1}dz$$

$$\ldots\ldots (3)$$

Which can be integrated, as was shown above, when $\frac{m}{n} + \frac{r}{s}$, is a whole number.

EXAMPLES.

1. $dy = (1 + x^2)^{\frac{1}{2}}x^3dx \ldots\ldots (4)$

Here, $\frac{m}{n} = \frac{4}{2} = 2$; hence, the second criterion is satisfied.

Comparing (2) with (4), we find,

$$a = 1,\ b = 1,\ n = 2,\ m = 4,\ r = 1,\ s = 2, \text{ and } z = (1 + x^2)^{\frac{1}{2}}$$

Hence,

$$\int(1 + x^2)^{\frac{1}{2}}x^3dx = \int(z^2 - 1)z^2dz = \frac{z^5}{5} - \frac{z^3}{3}$$

$$= \frac{(1 + x^2)^{\frac{5}{2}}}{5} - \frac{(1 + x^2)^{\frac{3}{2}}}{3} = \frac{(3x^2 - 2)(1 + x^2)^{\frac{3}{2}}}{15}.$$

2. $dy = \dfrac{dx}{x^4(1+x^2)^{\frac{1}{2}}} = dx(1+x^2)^{-\frac{1}{2}}x^{-4} \;.\;.\;.\;.\;.\; (5)$

Here, $\dfrac{m}{n} + \dfrac{r}{s} = -\dfrac{3}{2} - \dfrac{1}{2} = -2$; hence, the third criterion is satisfied.

Comparing (3) and (5), we find,

$a = 1,\ b = 1,\ n = 2,\ m = -3,\ r = -1,\ s = 2,$ and

$$z = \frac{\sqrt{1+x^2}}{x}.$$

Hence,

$$\int \frac{dx}{x^4(1+x^2)^{\frac{1}{2}}} = -\int (z^2 - 1)dz = z - \frac{z^3}{3} = \frac{3z - z^3}{3}$$

$$= \frac{(2x^2 - 1)(1+x^2)^{\frac{1}{2}}}{3x^3}.$$

Integration by Successive Reduction.

70. When an integral can be made to depend upon a simpler integral of the same form, it is evident that by successive repetitions of the process, we may ultimately arrive at a form that can be integrated by one of the fundamental formulas. This method of integration is called *the method by successive reduction*, and is effected by *formulas of reduction*, some of which it is now proposed to investigate.

1°. If in the expression for a binomial differential, we omit the factor A, represent the exponent of the parenthesis by p, and factor the result, we have,

$$(a + bx^n)^p x^{m-1}dx = x^{m-n}(a + bx^n)^p x^{n-1}dx \;.\;.\;.\;.\;.\; (1)$$

Let $u = x^{m-n}$, and $dv = (a + bx^n)^p x^{n-1}dx$.

Differentiating the first, and integrating the second by Formula (29), we have,

$$du = (m - n)x^{m-n-1}dx; \text{ and } v = \frac{(a + bx^n)^{p+1}}{bn(p + 1)};$$

But, $(a + bx^n)^{p+1} = (a + bx^n)^p\,(a + bx^n)$;

$$\therefore\ v = \frac{a(a + bx^n)^p}{bn(p + 1)} + \frac{(a + bx^n)^p x^n}{n(p + 1)}.$$

Substituting in Formula (28), we have,

$$\int(a + bx^n)^p x^{m-1}dx = \frac{(a + bx^n)^{p+1}x^{m-n}}{bn(p + 1)}$$
$$- \frac{a(m - n)}{bn(p + 1)}\int(a + bx^n)^p x^{m-n-1}dx$$
$$- \frac{(m - n)}{n(p + 1)}\int(a + bx^n)^p x^{m-1}dx.$$

Transposing the last term to the first member, and reducing, we have,

$$\frac{pn + m}{n(p + 1)}\int(a + bx^n)^p x^{m-1}dx = \frac{(a + bx^n)^{p+1}x^{m-n}}{bn(p + 1)}$$
$$- \frac{a(m - n)}{bn(p + 1)}\int(a + bx^n)^p x^{m-n-1}dx.$$

Dividing by the coefficient in the first member, we have,

$$\int(a + bx^n)^p x^{m-1}dx = \frac{(a + bx^n)^{p+1}x^{m-n}}{b(pn + m)}$$
$$- \frac{a(m - n)}{b(pn + m)}\int(a + bx^n)^p x^{m-n-1}dx \quad \ldots\ldots \quad [A]$$

Formula A enables us to reduce the exponent of the variable without the parenthesis, by the exponent of the variable within the parenthesis, at each application. It fails when $pn + m = 0$; but in this case, the third criterion, (Art. 69), is satisfied, and the expression may be integrated by a previous method.

2°. The expression $(a + bx^n)^p$ may be placed under the form,

$$(a + bx^n)^{p-1}(a + bx^n),$$

$$\text{or, } a(a + bx^n)^{p-1} + b(a + bx^n)^{p-1}x^n;$$

hence, we may write the equation,

$$\int (a + bx^n)^p x^{m-1} dx$$

$$= a\int (a + bx^n)^{p-1} x^{m-1} dx + b\int (a + bx^n)^{p-1} x^{m+n-1} dx \quad \ldots\ldots (2)$$

The last term of this equation can be reduced by Formula A. Replacing m by $m + n$, p by $p - 1$, and multiplying by b, that formula becomes,

$$b\int (a + bx^n)^{p-1} x^{m+n-1} dx$$

$$= \frac{(a + bx^n)^p x^m}{pn + m} - \frac{am}{pn + m}\int (a + bx^n)^{p-1} x^{m-1} dx.$$

Substituting in (2), and uniting similar terms, we have,

$$\int (a + bx^n)^p x^{m-1} dx$$

$$= \frac{(a + bx^n)^p x^m}{pn + m} + \frac{pna}{pn + m}\int (a + bx^n)^{p-1} x^{m-1} dx \ldots [B]$$

Formula B enables us to reduce the exponent of the parenthesis by 1 at each application. It fails in the same case as Formula A. When m and p are negative, we endeavor to increase instead of diminishing them. For this purpose we reverse Formulas A and B.

3°. Reversing Formula A, and reducing, we have,

$$\int (a + bx^n)^p x^{m-n-1} dx$$

$$= \frac{(a + bx^n)^{p+1} x^{m-n}}{a(m - n)} - \frac{b(pn + m)}{a(m - n)} \int (a + bx^n)^p x^{m-1} dx.$$

Replacing m by $-m + n$, we have,

$$\int (a + bx^n)^p x^{-m-1} dx = -\frac{(a + bx^n)^{p+1} x^{-m}}{am}$$

$$+ \frac{b(pn - m + n)}{am} \int (a + bx^n)^p x^{-m+n-1} dx \dots\dots [C]$$

4°. Reversing Formula B, and reducing, we have,

$$\int (a + bx^n)^{p-1} x^{m-1} dx$$

$$= -\frac{(a + bx^n)^p x^m}{pna} + \frac{pn + m}{pna} \int (a + bx^n)^p x^{m-1} dx.$$

In this, substituting $-p + 1$ for p, we have,

$$\int (a + bx^n)^{-p} x^{m-1} dx = \frac{(a + bx^n)^{-p+1} x^m}{an(p - 1)}$$

$$- \frac{m + n - np}{an(p - 1)} \int (a + bx^n)^{-p+1} x^{m-1} dx \dots\dots [D]$$

The mode of applying Formulas A, B, C, and D, will be illustrated by practical examples.

1. Let it be required to integrate the expression,

$$(r^2 - x^2)^{\frac{1}{2}} x^2 dx.$$

In Formula A, making

$$a = r^2,\ b = -1,\ n = 2,\ p = \tfrac{1}{2},\ \text{and } m = 3,$$

we find,

$$\int (r^2 - x^2)^{\frac{1}{2}} x^2 dx = -\frac{(r^2 - x^2)^{\frac{3}{2}} x}{4} + \frac{r^2}{4} \int (r^2 - x^2)^{\frac{1}{2}} dx \quad \ldots\ldots (3)$$

In Formula B, making

$$a = r^2,\ b = -1,\ n = 2,\ p = \tfrac{1}{2},\ \text{and } m = 1,$$

we find,

$$\int (r^2 - x^2)^{\frac{1}{2}} dx = \frac{(r^2 - x^2)^{\frac{1}{2}} x}{2} + \frac{r^2}{2} \int (r^2 - x^2)^{-\frac{1}{2}} dx \quad \ldots\ldots (4)$$

But $\int (r^2 - x^2)^{-\frac{1}{2}} dx = \int \frac{dx}{\sqrt{r^2 - x^2}} = \sin^{-1}\frac{x}{r}$.

Substituting this in (4), and that result in (3), we have, finally,

$$\int (r^2 - x^2)^{\frac{1}{2}} x^2 dx = -\frac{(r^2 - x^2)^{\frac{3}{2}} x}{4} + \frac{r^2 (r^2 - x^2)^{\frac{1}{2}} x}{8} + \frac{r^4}{8} \sin^{-1}\frac{x}{r}$$

2. Let $dy = \dfrac{dx}{x^3(x^2 - a^2)^{\frac{1}{2}}} = (-a^2 + x^2)^{-\frac{1}{2}}x^{-3}dx.$

In Formula C, making $a = -a^2$, $b = 1$, $n = 2$, $p = -\frac{1}{2}$, and $m = 2$, we have,

$$\int(-a^2 + x^2)^{-\frac{1}{2}}x^{-3}dx = +\frac{(x^2 - a^2)^{\frac{1}{2}}x^{-2}}{2a^2}$$

$$+\frac{1}{2a^2}\int(-a^2 + x^2)^{-\frac{1}{2}}x^{-1}dx \; . \; . \; . \; . \; . \; (5)$$

But by Formula (24),

$$\int(-a^2 + x^2)^{-\frac{1}{2}}x^{-1}dx = \int\frac{dx}{x\sqrt{x^2 - a^2}} = \frac{1}{a}\sec^{-1}\frac{x}{a}.$$

Substituting in (5), we have, for the value of y,

$$\int\frac{dx}{x^3(x^2 - a^2)^{\frac{1}{2}}} = \frac{(x^2 - a^2)^{\frac{1}{2}}}{2a^2x^2} + \frac{1}{2a^3}\sec^{-1}\frac{x}{a}.$$

3. Let $dy = \dfrac{dz}{(z^2 + 1)^2} = (z^2 + 1)^{-2}dz.$

In Formula D, making $x = z$, $a = 1$, $b = 1$, $n = 2$, $p = 2$, and $m = 1$, we have,

$$\int(z^2 + 1)^{-2}\,dz = \frac{(z^2 + 1)^{-1}z}{2} + \frac{1}{2}\int(z^2 + 1)^{-1}dz.$$

But,

$$\int(z^2 + 1)^{-1}dz = \int\frac{dz}{1 + z^2} = \tan^{-1}z.$$

Hence, we have, for the value of y,

$$\int \frac{dz}{(z^2 + 1)^2} = \frac{z}{2(z^2 + 1)} + \frac{1}{2} \tan^{-1} z.$$

This is the method of integration referred to under the Fourth Case of rational fractions. By a continued application of Formula D, any expression of the form, $\frac{dz}{(a^2 + z^2)^n}$, can be reduced to an integrable form.

The following additional examples can be reduced to integrable forms by the application of Formulas A, B, C, and D.

4. $dy = \frac{x^4 dx}{(a^2 - x^2)^{\frac{1}{2}}}$.

$$Ans.\ y = -\frac{x^3 (a^2 - x^2)^{\frac{1}{2}}}{4} - \frac{3a^2 x}{8}(a^2 - x^2)^{\frac{1}{2}} + \frac{3a^4}{8} \sin^{-1}\frac{x}{a}.$$

5. $dy = (a^2 - x^2)^{\frac{1}{2}} dx$.

$$Ans.\ y = \frac{1}{2} x (a^2 - x^2)^{\frac{1}{2}} + \frac{a^2}{2} \sin^{-1}\frac{x}{a}.$$

6. $dy = (1 + x^2)^{\frac{3}{2}} x^3 dx$. $\quad Ans.\ y = \frac{5x^2 - 2}{35} (1 + x^2)^{\frac{5}{2}}$.

7. $dy = \frac{x^3 dx}{(1 + x^2)^{\frac{3}{2}}}$. $\quad Ans.\ y = \frac{x^2 + 2}{\sqrt{1 + x^2}}$.

8. $dy = \frac{x^3 dx}{(1 + x^2)^{\frac{5}{2}}}$. $\quad Ans.\ y = -\frac{3x^2 + 2}{3(1 + x^2)^{\frac{3}{2}}}$.

9. $dy = \frac{x^4 dx}{(1 - x^2)^{\frac{3}{2}}}$. $\quad Ans.\ y = -\frac{x^3 - 3x}{2\sqrt{1 - x^2}} - \frac{3}{2} \sin^{-1} x$.

10. $dy = \frac{dx}{x^4(1 - x^2)^{\frac{1}{2}}}$. *Ans.* $y = -\frac{2x^2 + 1}{3x^3}(1 - x^2)^{\frac{1}{2}}$

5°. Let it be required to simplify the expression,

$$\frac{x^n dx}{\sqrt{2ax - x^2}}.$$

Changing the form and removing the factor x from under the radical sign, we have,

$$(2a - x)^{-\frac{1}{2}} x^{n - \frac{1}{2}} dx\,;$$

in formula A, making,

$$a = 2a,\ b = -1,\ n = 1,\ p = -\tfrac{1}{2}, \text{ and } m = n + \tfrac{1}{2},$$

and reducing, we have,

$$b(pn + m) = -n, \text{ and } a(m - n) = 2a(n - \tfrac{1}{2}) = a(2n-1);$$

hence,

$$\int \frac{x^n dx}{\sqrt{2ax - x^2}} = -\frac{x^{n-1}\sqrt{2ax - x^2}}{n}$$

$$+ \frac{(2n - 1)a}{n}\int \frac{x^{n-1}dx}{\sqrt{2ax - x^2}} \quad \ldots\ldots \quad [E]$$

Formula E enables us to reduce the exponent n by 1, at each application; if n is entire and positive, by a continued application of the formula we ultimately arrive at the form, $\int \frac{dx}{\sqrt{2ax-x^2}}$, which is equal to $\text{versin}^{-1}\frac{x}{a}$.

EXAMPLES.

1. $dy = \dfrac{xdx}{\sqrt{2ax - x^2}}$.

$$Ans.\ y = -\sqrt{2ax - x^2} + a \text{ versin}^{-1}\frac{x}{a}$$

2. $dy = \dfrac{x^2 dx}{\sqrt{2x - x^2}}$.

$$Ans.\ y = -\frac{x+3}{2}\sqrt{2x - x^2} + \frac{3}{2}\text{ versin}^{-1}x.$$

3. $dy = \dfrac{x^3 dx}{\sqrt{2x - x^2}}$.

$$Ans.\ y = -\frac{2x^2 + 5x + 15}{6}\sqrt{2x - x^2} + \frac{5}{2}\text{ versin}^{-1}x.$$

Certain Trinomial Differentials.

71. Every trinomial differential that can be reduced to the form

$$(a + bx \pm x^2)^{\frac{p}{2}} x^m dx \quad \ldots\ldots \quad (1)$$

can be made rational in terms of an auxiliary variable, and consequently can be integrated, when p and m are whole numbers.

When p is *even*, the expression is already rational; when p is odd, say of the form $2n + 1$, the expression becomes,

$$(a + bx \pm x^2)^n (a + bx \pm x^2)^{\frac{1}{2}} x^m dx \quad \ldots\ldots \quad (2)$$

in which the only irrational part is $\sqrt{a + bx \pm x^2}$. There may be two cases; *first*, when x^2 is positive, and *secondly*, when x^2 is negative.

1°. *When x^2 is positive.*

Assume,

$$\sqrt{a + bx + x^2} = z - x; \quad \therefore\ a + bx + x^2 = z^2 - 2zx + x$$

striking out the common term x^2, and solving, we have,

$$x = \frac{z^2 - a}{2z + b}; \quad \therefore\ dx = \frac{2(z^2 + bz + a)}{(2z + b)^2} dz,$$

and

$$\sqrt{a + bx + x^2} = \frac{z^2 + bz + a}{2z + b} \quad \ldots\ldots \quad (3)$$

Substituting these in (2), the result will be rational.

EXAMPLES.

1. $dy = \dfrac{dx}{\sqrt{1 + x + x^2}} = (1 + x + x^2)^{-\frac{1}{2}} dx \quad \ldots\ldots \quad (4)$

Comparing (4) with (1), we find, $a = 1$, and $b = 1$. Hence, from (3), we have,

$$\sqrt{1 + x + x^2} = \frac{z^2 + z + 1}{2z + 1}, \text{ and } dx = \frac{2(z^2 + z + 1)}{(2z + 1)^2} dz.$$

Substituting in (4), and making $z = x + \sqrt{1 + x + x^2}$, we have,

$$dy = \frac{2dz}{2z + 1};$$

$$\therefore\ y = l(2z + 1) = l(2x + 1 + 2\sqrt{1 + x + x^2})$$

2. $dy = \dfrac{dx}{\sqrt{x^2 - x - 1}}$.

Ans. $y = l(2x - 1 + 2\sqrt{x^2 - x - 1})$.

3. $dy = \frac{dx}{x\sqrt{1 + x + x^2}}$.

$$Ans.\ y = l\left(\frac{3x}{2 + x + 2\sqrt{1 + x + x^2}}\right).$$

2°. *When x^2 is negative.*

Let α and β be taken so as to satisfy the equation,

$$a + bx - x^2 = (x - \alpha)(\beta - x) \quad . \quad . \quad . \quad . \quad . \quad (5)$$

Assume

$$\sqrt{a + bx - x^2} = \sqrt{(x - \alpha)(\beta - x)} = (x - \alpha)z;$$

squaring and reducing, we have,

$$\beta - x = (x - \alpha)z^2; \text{ or, } z = \sqrt{\frac{\beta - x}{x - \alpha}};$$

hence,

$$x = \frac{\beta + \alpha z^2}{1 + z^2}; \quad dx = -\frac{2(\beta - \alpha)zdz}{(1 + z^2)^2};$$

and

$$\sqrt{a + bx - x^2} = \frac{(\beta - \alpha)z}{1 + z^2} \quad . \quad . \quad . \quad . \quad . \quad (6)$$

These values substituted in (1), will make it rational.

EXAMPLE.

Let $dy = \frac{dx}{\sqrt{2 - x - x^2}} = \frac{dx}{\sqrt{(x + 2)(1 - x)}}$;

$$\therefore \alpha = -2, \text{ and } \beta = 1 \quad . \quad . \quad . \quad . \quad . \quad (7)$$

From (6), we have,

$$dx = -\frac{6zdz}{(1+z^2)^2}, \text{ and } \sqrt{2-x-x^2} = \frac{3z}{(1+z^2)};$$

$$\text{also } z = \sqrt{\frac{1-x}{x+2}}$$

Substituting in (7), we find,

$$dy = -\frac{2dz}{1+z^2}; \therefore y = -2\tan^{-1}z = -2\tan^{-1}\sqrt{\frac{1-x}{x+2}}.$$

Integration by Series.

72. It is often convenient to develop a differential into a series, arranged according to the ascending powers of the variable, and then to integrate each term separately. This method sometimes enables us to find an approximate value for an integral, which cannot be found in any other manner; it also enables us to deduce many useful formulas.

EXAMPLES.

1. $dy = x^2(1-x^2)^{\frac{1}{2}}dx.$

Developing $(1-x^2)^{\frac{1}{2}}$ by the binomial formula, and multiplying each term by x^2dx, we have,

$$dy = (x^2 - \frac{x^4}{2} - \frac{x^6}{8} - \frac{x^8}{16} - \text{etc.})dx.$$

$$\therefore y = \frac{x^3}{3} - \frac{x^5}{10} - \frac{x^7}{56} - \frac{x^9}{144} - \text{etc.}$$

2. $dy = \dfrac{dx}{1+x^2} = dx(1+x^2)^{-1}.$

By Formula (14) y is equal to $\tan^{-1}x$. Developing by the binomial formula, we have,

$$(1 + x^2)^{-1} = 1 - x^2 + x^4 - x^6 + x^8 - x^{10} + \text{etc.}$$

Substituting, and integrating, we have the Formula,

$$\tan^{-1}x = x - \frac{x^3}{3} + \frac{x^5}{5} - \frac{x^7}{7} + \frac{x^9}{9} - \text{etc.}$$

3. $dy = \dfrac{dx}{1 + x} = dx(1 + x)^{-1}$

$$= (1 - x + x^2 - x^3 + \text{etc.})dx$$

$$\therefore\ y = l(1 + x) = x - \frac{x^2}{2} + \frac{x^3}{3} - \frac{x^4}{4} + \text{etc.}$$

4. $dy = \dfrac{dx}{\sqrt{1 - x^2}} = dx(1 - x^2)^{-\frac{1}{2}}.$

Ans. $y = \sin^{-1}x = x + \dfrac{x^3}{1.2.3} + \dfrac{3x^5}{2.4.5} + \dfrac{3.5x^7}{2.4.6.7} + \text{etc.}$

Integration of Transcendental Differentials.

73. A transcendental differential is one that is expressed in terms of some transcendental quantity. There are three principal classes; viz., *logarithmic*, *exponential*, and *circular*.

Logarithmic Differentials.

74. A large number of cases come under the general form,

$$dy = x^{m-1}(lx)^n dx \quad . \; . \; . \; . \; . \quad (1)$$

Assume,

$$u = (lx)^n, \text{ and } dv = x^{m-1}dx;$$

$$\therefore du = \frac{n(lx)^{n-1}dx}{x}, \text{ and } v = \frac{x^m}{m};$$

substituting in Formula (28), and reducing, we have,

$$y = \int x^{m-1}(lx)^n dx = \frac{x^m (lx)^n}{m} - \frac{n}{m}\int x^{m-1}(lx)^{n-1}dx \ldots [F]$$

Formula F is a formula of reduction; it reduces the exponent of (lx) by 1 at each application.

EXAMPLES.

1. $dy = x(lx)^3 dx$.

$$Ans.\ y = \frac{x^2}{2}(lx)^3 - \frac{3x^2}{4}(lx)^2 + \frac{3x^2}{4}(lx) - \frac{3x^2}{8}.$$

2. $dy = x^3(lx)dx$. $\qquad Ans.\ y = \frac{x^4(lx)}{4} - \frac{x^4}{16}.$

3. $dy = x^3(lx)^3 dx$.

$$Ans.\ y = \frac{x^4}{4}\left((lx)^3 - \frac{3}{4}(lx)^2 + \frac{3}{8}(lx) - \frac{3}{32}\right).$$

Reversing Formula F, and reducing, we have,

$$\int x^{m-1}(lx)^{n-1}dx = \frac{x^m(lx)^n}{n} - \frac{m}{n}\int x^{m-1}(lx)^n dx.$$

Replacing $n - 1$ by $- n$, and reducing, we have,

$$\int \frac{x^{m-1}dx}{(lx)^n} = -\frac{x^m}{(n-1)(lx)^{n-1}} + \frac{m}{n-1}\int \frac{x^{m-1}dx}{(lx)^{n-1}} \quad \ldots\ldots [G]$$

The continued application of formula G gives as a final integral, the expression $\int \frac{x^{m-1}dx}{(lx)}$, which can be integrated by series. It fails when $n = 1$. Let $m = 0$, and $n = 1$; we then have,

$$\int \frac{dx}{xlx} = \int \frac{d(lx)}{lx} = l(lx) = l^2x.$$

Exponential Differentials.

75. Let it be required to integrate an expression of the form,

$$dy = x^m a^x dx \quad . \quad . \quad . \quad . \quad . \quad (1)$$

Assume,

$$u = x^m, \text{ and } dv = a^x dx; \ \therefore \ du = mx^{m-1}dx, \text{ and } v = \frac{a^x}{la}.$$

Substituting in Formula (28),

$$\int x^m a^x dx = \frac{x^m a^x}{la} - \frac{m}{la}\int x^{m-1}a^x dx \quad . \quad . \quad . \quad . \quad . \quad [H]$$

When $m < 1$, we have,

$$\int \frac{a^x dx}{x^n} = -\frac{a^x}{(m-1)x^{m-1}} + \frac{la}{m-1}\int \frac{a^x dx}{x^{m-1}} \quad . \quad . \quad . \quad . \quad . \quad [I]$$

EXAMPLES.

1. $dy = xa^x dx$. *Ans.* $y = \frac{a^x}{la}\left(x - \frac{1}{la}\right)$.

2. $dy = x^3 a^x dx$.

$$Ans.\ y = \frac{a^x}{la}\left(x^3 - \frac{3x^2}{la} + \frac{6x}{(la)^2} - \frac{6}{(la)^3}\right).$$

3. $dy = \frac{e^x dx}{x^2}$. *Ans.* $y = -\frac{e^x}{x} + \int \frac{e^x dx}{x}$.

But by means of McLaurin's Formula, we find,

$$e^x = 1 + x + \frac{x^2}{1.2} + \frac{x^3}{1.2.3} + \frac{x^4}{1.2.3.4} + \text{etc.};$$

hence,

$$\int \frac{e^x dx}{x} = \int \left(\frac{dx}{x} + dx + \frac{xdx}{1.2} + \text{etc.}\right)$$

$$= lx + x + \frac{x^2}{1.2.2} + \frac{x^3}{1.2.3.3} + \text{etc.},$$

which, being substituted above, gives,

$$y = -\frac{e^x}{x} + lx + x + \frac{x^2}{1.2.2} + \frac{x^3}{1.2.3.3} + \text{etc.}$$

4. $dy = x^2 e^x dx$. *Ans.* $y = e^x(x^2 - 2x + 2)$.

Circular Differentials.

76. Circular differentials may be integrated by the methods of *transformation* and *successive reduction*, or they may be reduced to algebraic forms by making $\sin x = z$.

EXAMPLES.

1. Let $dy = \frac{dx}{\sin x}$.

From trigonometry, we have,

$$\sin x = 2\sin\tfrac{1}{2}x\cos\tfrac{1}{2}x = 2\tan\tfrac{1}{2}x\cos^2\tfrac{1}{2}x,$$

or,

$$\frac{dx}{\sin x} = \frac{dx}{2\tan\frac{1}{2}x\cos^2\frac{1}{2}x} = \frac{d(\tan\frac{1}{2}x)}{\tan\frac{1}{2}x}.$$

Hence,

$$\int\frac{dx}{\sin x} = l(\tan\tfrac{1}{2}x) \quad . \; . \; . \; . \; . \quad (1)$$

2. Let $dy = \frac{dx}{\cos x}$.

If we make x equal to $90° - x$ in the preceding formula, and reduce, we have,

$$\int\frac{dx}{\cos x} = -\,l[\tan(45° - \tfrac{1}{2}x)] \quad . \; . \; . \; . \; . \quad (2)$$

3. Let $dy = \frac{dx}{\tan x}$.

From trigonometry, we have,

$$\tan x = \frac{\sin x}{\cos x}; \text{ or, } \frac{dx}{\tan x} = \frac{dx\cos x}{\sin x} = \frac{d(\sin x)}{\sin x}.$$

Hence,

$$\int\frac{dx}{\tan x} = l(\sin x) \quad . \; . \; . \; . \; . \quad (3)$$

4. Let $dy = \frac{dx}{\cot x}$.

Making x equal to $90° - x$ in (3), and reducing, we have,

$$\int \frac{dx}{\cot x} = -\, l(\cos x) \quad . \; . \; . \; . \; . \quad (4)$$

5. Let $dy = \frac{dx}{\sin x \cos x}$.

From trigonometry, we have,

$$\sin x \cos x = \tfrac{1}{2}\sin 2x;$$

we also have,

$$dx = \tfrac{1}{2}d(2x);$$

hence,

$$\frac{dx}{\sin x \cos x} = \frac{d(2x)}{\sin 2x}.$$

Applying Formula (1), we have,

$$\int \frac{dx}{\sin x \cos x} = l(\tan x) \quad . \; . \; . \; . \; . \quad (5)$$

Let us have the expression,

$$dy = \sin^m x \cos^n x dx \quad . \; . \; . \; . \; . \quad (6)$$

Making $\sin x = z$, whence $\cos x = \sqrt{1 - z^2}$,

and $dx = \frac{dz}{\sqrt{1 - z^2}}$,

we have,

$$dy = z^m(1 - z^2)^{\frac{n-1}{2}} dz \quad . \; . \; . \; . \; . \quad (7)$$

Form (7) can always be integrated when m and n are whole numbers, because it will then satisfy one of the *three* criterions (Art. 69).

6. $dy = \sin^2 x \cos^3 x dx$.

Comparing with (6), we find $m = 2$, and $n = 3$; hence, from (7), we have,

$$dy = (1 - z^2)z^2 dz; \quad \therefore\ y = \frac{z^3}{3} - \frac{z^5}{5} = \sin^3 x(\tfrac{1}{3} - \tfrac{1}{5}\sin^2 x).$$

7. Let $dy = \sin^3 x dx$.

Here $m = 3$, and $n = 0$; and $dy = (1 - z^2)^{-\frac{1}{2}} z^3 dz$; making $x = z$, $a = 1$, $b = -1$, $n = 2$, $p = -\frac{1}{2}$, and $m = 4$, in Formula A, we have,

$$\int (1 - z^2)^{-\frac{1}{2}} z^3 dz = -\frac{(1 - z^2)^{\frac{1}{2}} z^2}{3} + \frac{2}{3}\int (1 - z^2)^{-\frac{1}{2}} z dz.$$

But from Formula (29),

$$\int (1 - z^2)^{-\frac{1}{2}} z dz = -(1 - z^2)^{\frac{1}{2}};$$

hence,

$$y = -\frac{1}{3}(1 - z^2)^{\frac{1}{2}} z^2 - \frac{2}{3}(1 - z^2)^{\frac{1}{2}} = -\frac{1}{3}\cos x(\sin^2 x + 2).$$

8. Let $dy = \cos^2 x dx = dz\sqrt{1 - z^2}$.

Applying Formula (B), we find,

$$\int (1 - z^2)^{\frac{1}{2}} dz = \frac{1}{2}(1 - z^2)^{\frac{1}{2}} z + \tfrac{1}{2}\int \frac{dz}{(1 - z^2)^{\frac{1}{2}}}.$$

But,

$$\int \frac{dz}{(1 - z^2)^{\frac{1}{2}}} = \int dx = x.$$

Hence,

$$y = \frac{1}{2}(\cos x \sin x + x).$$

9. Let $dy = \frac{dx}{\sin^3 x} = \frac{dz}{z^3\sqrt{1 - z^2}} = (1 - z^2)^{-\frac{1}{2}} z^{-3} dz.$

Applying Formula C, we have,

$$\int (1 - z^2)^{-\frac{1}{2}} z^{-3} dz = -\frac{(1 - z^2)^{\frac{1}{2}} z^{-2}}{2} + \frac{1}{2}\int (1 - z^2)^{-\frac{1}{2}} z^{-1} dz.$$

But,

$$\int (1 - z^2)^{-\frac{1}{2}} z^{-1} dz = \int \frac{dz}{z\sqrt{1 - z^2}} = \int \frac{dx}{\sin x} = l(\tan \tfrac{1}{2}x).$$

Hence,

$$y = -\frac{\cos x}{2\sin^2 x} + \frac{1}{2} l(\tan \tfrac{1}{2}x).$$

10. Let $dy = \frac{dx}{\sin x \cos^3 x} = \frac{dz}{z(1 - z^2)^2} = (1 - z^2)^{-2} z^{-1} dz.$

By Formula D, we have,

$$\int (1 - z^2)^{-2} z^{-1} dz = \frac{1}{2}(1 - z^2)^{-1} + \int (1 - z^2)^{-1} z^{-1} dz.$$

But,

$$\int (1 - z^2)^{-1} z^{-1} dz = \int \frac{dz}{z\sqrt{1 - z^2}\sqrt{1 - z^2}}$$

$$= \int \frac{dx}{\sin x \cos x} = l(\tan x).$$

Hence,

$$y = \frac{1}{2\cos^2 x} + l(\tan x).$$

11. Let $dy = \tan^4 x dx = (1 - z^2)^{-\frac{5}{2}} z^4 dz$,

$$y = \tfrac{1}{3}\tan^3 x - \tan x + x.$$

EXAMPLES IN SUCCESSIVE INTEGRATION.

12. $d^2y = \sin x \cos^2 x dx^2 = \sin x (d\sin x)^2$.

Making $\sin x = z$, whence, $d^2y = z(dz)^2$, we have,

$$\frac{dy}{dz} = \frac{z^2}{2} + C, \text{ and } y = \frac{z^3}{6} + Cz + C',$$

or,

$$y = \frac{\sin^3 x}{6} + C\sin x + C'.$$

13. $d^2y = \cos x \sin^2 x dx^2 = \cos x (d\cos x)^2$;

$$\therefore \frac{dy}{d(\cos x)} = \frac{1}{2}\cos^2 x + C,$$

and

$$y = \frac{1}{6}\cos^3 x + C\cos x + C'.$$

14. $d^2y = -\dfrac{dx^2}{x^2}$;

$$\therefore \frac{dy}{dx} = \frac{1}{x} + C; \text{ or, } y = lx + Cx + C'$$

Integration of Differential Functions of two Variables.

77. A total differential of a function of two variables is of the form,

$$dz = Pdx + Qdy \;.\;.\;.\;.\;.\; (1)$$

In which P and Q are functions of x and y; but differential expressions of that form are not necessarily total differentials. That they may be so, they must (Eq. 5, Art. 26), satisfy the condition,

$$\frac{dP}{dy} = \frac{dQ}{dx} \quad \ldots\ldots \quad (2)$$

When an expression of the form (1) is to be integrated, we first apply the test expressed by (2); if that be satisfied, the expression is *integrable.* To perform the integration, integrate one of the partial differentials with reference to its corresponding variable, that is, as though the other variable were constant, and add such a function of that variable as will satisfy equation (1). Thus, to integrate expression (1), we have,

$$z = \int P dx + R \quad \ldots\ldots \quad (3)$$

in which R is a function of y alone. The value of R may be found by differentiating the second member of (3) with respect to y, and placing its partial differential coefficient equal to Q. Hence, we have,

$$\frac{d\int P dx}{dy} + \frac{dR}{dy} = Q.$$

Transposing, multiplying by dy, and integrating with respect to y, that is, as though x were constant, we find,

$$R = \int \left(Q - \frac{d\int P dx}{dy} \right) dy.$$

Substituting in (3), we find,

$$z = \int Pdx + \int \left(Q - \frac{d\int Pdx}{dy}\right)dy \; . \; . \; . \; . \; . \; (4)$$

EXAMPLES.

1. $dz = 3x^2y^2dx + 2x^3ydy.$

Here $P = 3x^2y^2$, and $Q = 2x^3y$, and (2) is satisfied. Integrating by Formula (4), we have,

$$z = x^3y^2 + \int (2x^3y - 2x^3y)dy = x^3y^2 + C.$$

2. $dz = ydx + xdy.$

The test (2) is satisfied. Integrating by (4), we have,

$$z = yx + \int (xdy - xdy)dy = yx + C.$$

3. $dz = \frac{dx}{y} + \left(2y - \frac{x}{y^2}\right)dy.$

By Formula (4), we have,

$$z = \frac{x}{y} + \int \left(2y - \frac{x}{y^2} + \frac{x}{y^2}\right)dy = \frac{x}{y} + y^2 + C.$$

4. $dz = (6xy - y^2)dx + (3x^2 - 2xy)dy.$

$$z = 3x^2y - y^2x + \int (3x^2 - 2xy - 3x^2 + 2xy)dy$$
$$= 3x^2y - y^2x + C.$$

PART IV.

APPLICATIONS OF THE INTEGRAL CALCULUS.

I. Lengths of Plane Curves.

Rectification.

78. *Rectification* is the operation of finding an expression for the length of a curve. This may be done by finding an element of the length, as explained in Art. 51, and then integrating it between proper limits. If we apply the integral sign to Equation (1), Art. 51, we have,

$$L = \int \sqrt{dx^2 + dy^2} \quad . \; . \; . \; . \; . \; (1)$$

1°. *To rectify the semi-cubic parabola, whose equation is,*

$$y^3 = a^2x^2.$$

Differentiating and substituting in (1), we have,

$$L = \frac{1}{2a}\int (4a^2 + 9y)^{\frac{1}{2}}dy \quad . \; . \; . \; . \; . \; (2)$$

Integrating by Formula (29), we have,

$$L = \frac{1}{27a}(4a^2 + 9y)^{\frac{3}{2}} + C \quad . \; . \; . \; . \; . \; (3)$$

As explained in Art. 60, this integral is indefinite, and

expresses the length of an arc from any ordinate up to any other ordinate.

If we estimate the length from a point whose ordinate is 0, the length at that point is 0, and, from (3), making L, and $y = 0$, we have,

$$0 = \frac{1}{27a}(4a^2)^{\frac{3}{2}} + C; \therefore C = -\frac{8a^2}{27}.$$

Substituting this value of C in (3), and denoting the corresponding value of L by L', we have,

$$L' = \frac{1}{27a}(4a^2 + 9y)^{\frac{3}{2}} - \frac{8a^2}{27} \quad . \quad . \quad . \quad . \quad . \quad (4)$$

This is a *particular* integral, and it expresses the length of the arc from the particular ordinate 0, up to any ordinate y.

If we wish to find the length from the particular ordinate 0, up to an ordinate b, we make $y = b$, in (4). Doing so, and denoting the corresponding value of L' by L'', we have,

$$L'' = \frac{1}{27a}(4a^2 + 9b)^{\frac{3}{2}} - \frac{8a^2}{27} \quad . \quad . \quad . \quad . \quad . \quad (5)$$

This is the *definite* integral, and expresses the length of the arc from the ordinate 0, up to the ordinate b.

In this case the *limits* are 0 and b; the integral may be found otherwise, as follows:

Making $y = 0$, in (3), and denote the corresponding value of L by L_0, we have,

$$L_0 = \frac{1}{27a}(4a^2)^{\frac{3}{2}} + C \quad . \quad . \quad . \quad . \quad . \quad (6)$$

Making $y = b$, in (3), and denoting the corresponding value of L by L_b, we have,

$$L_b = \frac{1}{27a}(4a^2 + 9b)^{\frac{3}{2}} + C \dots . (7)$$

Equation (6) expresses the length of the curve from any point up to the point whose ordinate is 0, and (7) expresses the length of the arc from the same point up to a point whose ordinate is b; the first taken from the second will therefore be the *definite* integral, and will express, as before, the length of the arc from the point whose ordinate is 0, up to the point whose ordinate is b. Making the subtraction, and adopting the notation explained in Art. 60, we have,

$$L'' = \frac{1}{2a}\int_0^b (4a^2 + 9y)^{\frac{1}{2}} dy = \frac{1}{27a}(4a^2 + 9b)^{\frac{3}{2}} - \frac{8a^2}{27} \dots (8)$$

The same result as found in (5). In this case the *initial* abscissa is 0 and the *terminal* abscissa is $\frac{b^{\frac{3}{2}}}{a}$.

2°. *To rectify the cycloid, whose differential equation* (Art. 56), *is*

$$dx = \pm \frac{ydy}{\sqrt{2ry - y^2}} \dots . . (9)$$

Substituting, in (1), and reducing, we have,

$$L = \sqrt{2r}\int (2r - y)^{-\frac{1}{2}} dy \dots . . (10)$$

Integrating between the limits 0 and $2r$ Formula (29), we have,

$$L'' = \sqrt{2r}\int_0^{2r} (2r - y)^{-\frac{1}{2}} dy = 4r \dots . . . (11)$$

This is the expression for *half* of one branch. Hence, the whole branch is equal to eight times the radius of the generating circle.

3°. *To rectify the circle, whose equation is,*

$$y = (r^2 - x^2)^{\frac{1}{2}}.$$

Differentiating, substituting in (1), and developing by the binomial formula, we have,

$$L = r\int(r^2 - x^2)^{-\frac{1}{2}}dx = \int\left(dx + \frac{1}{2r^2} \cdot x^2dx + \frac{1.3}{2.4r^4} \cdot x^4dx + \frac{1.3.5}{2.4.6r^6}x^6dx + \text{etc.}\right) \quad . \ . \ . \ . \ . \ (12)$$

Performing the integration, making $r = 1$, and commencing the arc at the point whose abscissa is 0, that is, at the upper extremity of the vertical diameter, we have,

$$L' = x + \frac{1}{3.2}x^3 + \frac{1.3}{5.2.4}x^5 + \frac{1.3.5}{7.2.4.6}x^7 + \text{etc.} \quad . \ . \ . \ . \ . \ (13)$$

Formula (13) may be used for finding the length of an arc, but the series is not very converging, and therefore the process is tedious. We know, when $x = \frac{1}{2}$, that L' is equal to an arc of 30°, that is, to $\frac{1}{6}\pi$. Making $L' = \frac{1}{6}\pi$, and $x = \frac{1}{2}$, we have,

$$\frac{1}{6}\pi = \frac{1}{2} + \frac{1}{2^3.2.3} + \frac{1.3}{2^5 2.4.5} + \frac{1.3.5}{2^7.2.4.6.7} + \text{etc.}, = .5235987;$$

$$\therefore \ \pi = 3.1415 \quad . \ . \ . \ . \ . \ (14)$$

4°. A better method for deducing the value of π, is to find the length of the arc, in terms of its tangent.

If $L = \tan^{-1}x$, we have (Art. 18),

$$dL = \frac{dx}{1 + x^2} = (1 + x^2)^{-1}dx.$$

Developing by the binomial formula, we have,

$$dL = dx(1 - x^2 + x^4 - x^6 + x^8 - \text{etc.}) \quad . \quad . \quad . \quad . \quad . \quad (15)$$

Integrating, and determining C, as before, we have,

$$L' = x - \frac{x^3}{3} + \frac{x^5}{5} - \frac{x^7}{7} + \frac{x^9}{9} - \text{etc.} \quad . \quad . \quad . \quad . \quad . \quad . \quad (16)$$

To render (16) converging, x must be made small. Assume the trigonometrical formula,

$$\tan(a + b) = \frac{\tan a + \tan b}{1 - \tan a \tan b},$$

and let $a = \tan^{-1}m$, $b = \tan^{-1}n$, and $a + b = \tan^{-1}z$; then will

$$z = \frac{m + n}{1 - mn}; \text{ or, } n = \frac{z - m}{mz + 1} \quad . \quad . \quad . \quad . \quad . \quad (17)$$

If $z = 1$, and $m = \frac{1}{5}$, we have, $n = \frac{2}{3}$;

$$\therefore \tan^{-1} 1 = \tan^{-1}\frac{1}{5} + \tan^{-1}\frac{2}{3}.$$

If $z = \frac{2}{3}$, and $m = \frac{1}{5}$, we have, $n = \frac{7}{17}$;

$$\therefore \tan^{-1}\frac{2}{3} = \tan^{-1}\frac{1}{5} + \tan^{-1}\frac{7}{17}.$$

If $z = \frac{7}{17}$, and $m = \frac{1}{5}$, we have, $n = \frac{9}{46}$;

$$\therefore \tan^{-1}\frac{7}{17} = \tan^{-1}\frac{1}{5} + \tan^{-1}\frac{9}{46}.$$

If $z = \frac{9}{46}$, and $m = \frac{1}{5}$, we have, $n = -\frac{1}{239}$;

$$\therefore \tan^{-1}\frac{9}{46} = \tan^{-1}\frac{1}{5} - \tan^{-1}\frac{1}{239}.$$

Hence, by successive substitution, we find,

$$\frac{\pi}{4} = \tan^{-1} 1 = 4\tan^{-1}\frac{1}{5} - \tan^{-1}\frac{1}{239} \quad . \quad . \quad . \quad . \quad . \quad (18)$$

If we make $x = \frac{1}{5}$, in Equation (16), we find the value of $\tan^{-1}\frac{1}{5}$, and in like manner, we find the $\tan^{-1}\frac{1}{239}$; substituting them in (18), and reducing, we have the value of π. With very little labor, the value of π may be found to 8 or 10 places of decimals.

II. Areas of Plane Curves.

Quadrature.

79. *Quadrature* is the operation of finding an expression for the area of a portion of a plane bounded by a curve, the axis of abscissas, and any two ordinates. This is done by finding an expression for the elementary area, as explained in Art. 51, and then integrating the result between proper limits. Applying the sign of integration to Formula (2), Art. 51, we have,

$$A = \int y dx \quad . \quad . \quad . \quad . \quad . \quad (1)$$

1°. *To find the area of a parabola, whose equation is,*

$$y^2 = 2px.$$

Finding the value of y, and substituting in (1), we have,

$$A = \sqrt{2p}\int x^{\frac{1}{2}} dx = \frac{2}{3}\sqrt{2p} \,.\, x^{\frac{3}{2}} + C \quad . \quad . \quad . \quad . \quad . \quad (2)$$

If we commence the area at the ordinate 0, we have, $C = 0$; hence,

$$A' = \frac{2}{3}\sqrt{2p}\,.\,x^{\frac{3}{2}} = \frac{2}{3}yx\,.\,.\,.\,.\,.\,(3)$$

That is, the area of any portion of a parabola, reckoned from the vertex, is two-thirds the rectangle of its terminal co-ordinates.

2°. *To find the area of a circle, whose equation is,*

$$y = \sqrt{r^2 - x^2}.$$

Substituting in (1), we find,

$$A = \int (r^2 - x^2)^{\frac{1}{2}} dx\,.\,.\,.\,.\,.\,(4)$$

Reducing by Formula B, and integrating the last term by Formula (20), we have,

$$A = \frac{x(r^2 - x^2)^{\frac{1}{2}}}{2} + \frac{r^2}{2}\sin^{-1}\frac{x}{r} + C\,.\,.\,.\,.\,.\,(5)$$

Taking the integral between the limits, $x = 0$ and $x = r$, which gives the area of a quadrant, and remembering that $\sin^{-1} 1 - \sin^{-1} 0 = \frac{1}{2}\pi$, we have,

$$A'' = \frac{r^2}{2}[\sin^{-1} 1 - \sin^{-1} 0] = \frac{\pi r^2}{4}; \therefore \text{ Area} = \pi r^2\,.\,.\,.\,(6)$$

3°. *To find the area of the ellipse, whose equation is,*

$$y = \frac{b}{a}(a^2 - x^2)^{\frac{1}{2}}.$$

Substituting in (1), and integrating, as in the last example, we have,

$$A = \frac{b}{a}\left[\frac{x(a^2 - x^2)^{\frac{1}{2}}}{2} + \frac{a^2}{2}\sin^{-1}\frac{x}{a}\right] + C \quad . . . . . (7)$$

Integrating from $x = 0$ to $x = a$, and multiplying by 4, we have, for the area of the ellipse,

$$4A'' = \pi ab \quad . . . . . (8)$$

4°. *To find the area of the cycloid, whose differential equation is,*

$$dx = \pm \frac{ydy}{\sqrt{2ry - y^2}}.$$

Substituting in (1), we find,

$$A = \int \frac{y^2 dy}{\sqrt{2ry - y^2}} \quad . . . . . (9)$$

Applying Formula E *twice,* and Formula (26) once, we find,

$$A = -\frac{y\sqrt{2ry - y^2}}{2}$$

$$+ \frac{3r}{2}\left[-\sqrt{2ry - y^2} + r\text{versin}^{-1}\frac{y}{r}\right] + C.$$

Taking the integral between the limits $y = 0$, and $y = 2r$, and multiplying by 2, we find, for the area of the cycloid,

$$2A'' = 3\pi r^2 \quad . . . . . (10)$$

That is, the area between one branch and the directing line is three times the area of the generating circle.

5°. *To find the area of the logarithmic curve, whose equation is,* $y = lx$.

Substituting in (1), we have,

$$A = \int (lx)dx \;.\;.\;.\;.\; (11)$$

Making $m = 1$ and $n = 1$, in Formula F, and reducing, we have,

$$A = xlx - x + C \;.\;.\;.\;.\;.\; (12)$$

If we commence the arc at the ordinate corresponding to $x = 1$, we have, $C = 1$, and

$$A' = x(lx - 1) + 1 \;.\;.\;.\;.\;.\; (13)$$

6°. *To find the area of a rectangular hyperbola, bounded by the curve, one asymptote, and any two ordinates to that asymptote.*

Assume the equation,

$$xy = m, \text{ and make } m = 1, \text{ whence } y = \frac{1}{x}.$$

Substituting in (1), we have,

$$A = \int \frac{dx}{x} = lx + C \;.\;.\;.\;.\;.\; (14)$$

Commencing the area from the ordinate through the vertex, where $x = 1$, we have, $C = 0$; hence,

$$A' = lx \;.\;.\;.\;.\;.\; (15)$$

That is, the area commencing from the ordinate through the vertex is the Napierian logarithm of the terminal abscissa. Had we not made $m = 1$, we should, in like manner, have found

$$A' = mlx.$$

In which the area is equal to the logarithm in a system whose modulus is m.

III. Areas of Surfaces of Revolution.

Surfaces Generated by the Revolution of Plane Curves.

80. The area of a portion of a surface of revolution, bounded by two planes perpendicular to its axis, is determined by finding an expression for an elementary zone, as explained in Art. 51, and integrating the result between proper limits. Applying the sign of integration to Formula (3), Art. 51, we have,

$$S = \int 2\pi y \sqrt{dx^2 + dy^2} \quad . \; . \; . \; . \; . \quad (1)$$

1°. *To find the surface of a sphere, the equation of the generating circle being,* $y = (r^2 - x^2)^{\frac{1}{2}}$.

Differentiating, we have, $dy = -(r^2 - x^2)^{-\frac{1}{2}} x dx$; substituting the values of y and dy in (1), and integrating from $-r$ to $+r$, we have,

$$S'' = 2\pi r \int_{-r}^{r} dx = 4\pi r^2 \quad . \; . \; . \; . \; . \quad (2)$$

Hence, the area of the surface of a sphere is equal to four great circles, or to two-thirds the surface of the circumscribed cylinder.

2°. *To find the surface of a right cone.*

The equation of the generating line, the vertex of the cone being at the origin, is

$$y = ax; \quad \therefore \; dy = adx.$$

Substituting in (1), and reducing, we have,

$$S = 2\pi a\sqrt{1 + a^2}\int x dx = \pi a x^2 \sqrt{1 + a^2} + C \; . \; . \; . \; . \; . \; (3)$$

If the initial plane pass through the vertex, we have, for that plane, $S = 0$, and $x = 0$, whence $C = 0$; hence,

$$S' = \pi a x^2 \sqrt{1 + a^2} = \pi y \times x\sqrt{1 + a^2} \; . \; . \; . \; . \; . \; (4)$$

But a is the tangent of the semi-angle of the cone, and consequently $x\sqrt{1 + a^2}$ is the slant height; $2\pi y$ is the circumference of the cone's base; hence, the convex surface is equal to half the circumference of the base into the slant height.

3°. *To find the surface of the paraboloid of revolution.*

Assume the equation $y^2 = 2px$, whence,

$$y = \sqrt{2px}, \text{ and } dy^2 = \frac{p}{2x} dx^2.$$

Substituting in (1), and reducing, we have,

$$S = 2\pi \int (p^2 + 2px)^{\frac{1}{2}} dx = \frac{2\pi}{3p} (p^2 + 2px)^{\frac{3}{2}} + C \; . \; . \; . \; . \; (5)$$

If the initial plane pass through the vertex, we have,

$$0 = \frac{2\pi p^2}{3} + C; \;\; \therefore \; C = -\frac{2\pi p^2}{3}.$$

Which, in (5), gives,

$$S' = \frac{2\pi}{3}\left[\frac{1}{p} (p^2 + 2px)^{\frac{3}{2}} - p^2\right] \; . \; . \; . \; . \; . \; (6)$$

4°. *To find the surface generated by revolving one* ***branch*** *of a cycloid about its base.*

Assuming the equation, $dx = \pm \frac{ydy}{\sqrt{2ry - y^2}}$, and substituting in (1), we have,

$$S = 2\pi\sqrt{2r}\int(2r - y)^{-\frac{1}{2}}ydy \quad . \; . \; . \; . \; . \quad (7)$$

Applying Formula A, and integrating the last term by Formula (29), we have,

$$S = -2\pi\sqrt{2r}\left[\frac{2}{3}y(2r - y)^{\frac{1}{2}} + \frac{8r}{3}(2r - y)^{\frac{1}{2}}\right] + C.$$

Integrating between the limits $y = 0$ and $y = 2r$, we have,

$$S'' = \frac{32}{3}\pi r^2 \quad . \; . \; . \; . \; . \quad (8)$$

This is the surface generated by half of one branch. Hence, to find the whole surface, we multiply by 2. This gives

$$2S'' = \frac{64}{3}\pi r^2.$$

IV. Volumes of Solids of Revolution.

Cubature.

81. *Cubature* is the operation of finding the volume of a solid. When this is bounded by a surface of revolution and two planes perpendicular to its axis, the volume may be ascertained by finding the volume of an element bounded by two such planes infinitely near to each other, as explained in Art. 51, and then integrating the result between proper limits. Applying the sign of integration to Formula (4), Art. 51, we have,

$$V = \pi\int y^2 dx \quad . \; . \; . \; . \; . \quad (1)$$

1°. *To find the volume of a sphere.*

Assume the equation, $y^2 = r^2 - x^2$, and combine it with (1); we have,

$$V = \pi\int (r^2 dx - x^2 dx) = \pi\left[r^2 x - \frac{x^3}{3}\right] + C.$$

Taking the integral between the limits, $x = -r$, and $x = +r$, we find,

$$V'' = \pi\left[2r^3 - \frac{2r^3}{3}\right] = \frac{4}{3}\pi r^3 = 4\pi r^2 \times \frac{1}{3}r \quad . \; . \; . \; . \; . \quad (2)$$

That is, *the volume is equal to the surface by one-third the radius.*

2°. *To find the volume of a spheroid of revolution.*

There are two species of spheroids of revolution.

1st. The *prolate spheroid,* generated by revolving an ellipse about its transverse axis.

2dly. The *oblate spheroid,* generated by revolving an ellipse about its conjugate axis.

1st. *The prolate spheroid.* In this case the equation of the meridian curve is, $y^2 = \frac{b^2}{a^2}(a^2 - x^2)$. Substituting this in (1), and integrating between the limits $x = -a$, and $x = +a$, we have,

$$V'' = \pi\frac{b^2}{a^2}\int_{-a}^{+a}(a^2 - x^2)dx = \frac{4}{3}\pi b^2 a = \frac{2}{3}\pi b^2 \times 2a \; . \; . \; . \; . \; . \; (3)$$

That is, *the volume is equal to two-thirds of the circumscribing cylinder.*

2dly. *The oblate spheroid.* In this case, if the conjugate axis coincide with the axis of x, the equation of the meridian curve is, $y^2 = \frac{a^2}{b^2}(b^2 - x^2)$. Substituting in (1), and integrating from $-b$ to $+b$, we have,

$$V'' = \frac{4}{3}\pi a^2 b = \frac{2}{3}\pi a^2 \times 2b \; . \; . \; . \; . \; . \; (4)$$

Hence, as before, we have, *the volume equal to two-thirds the circumscribing cylinder.*

In both cases, if $a = b = r$, we have,

$$V'' = \frac{4}{3}\pi r^3 \; . \; . \; . \; . \; . \; (5)$$

This hypothesis causes the ellipsoids to merge into the sphere.

3°. *To find the volume of a paraboloid of revolution.*

The equation of the meridian curve is, $y^2 = 2px$. Hence, from (1), we have,

$$V = 2\pi p\int x dx = \pi p x^2 + C \; . \; . \; . \; . \; . \; (6)$$

If the initial plane pass through the vertex, we have $C = 0$, and

$$V' = \pi p x^2 = \pi y^2 \times \tfrac{1}{2}x \; . \; . \quad . \; . \; (7)$$

That is, *the volume is equal to half the cylinder that has the same base and the same altitude.*

4°. *To find the volume generated by revolving one branch of the cycloid about its base.*

The differential equation of the meridian curve is,

$$dx = \frac{ydy}{\sqrt{2ry - y^2}};$$

hence, from (1), we have,

$$V = \pi \int \frac{y^3 dy}{\sqrt{2ry - y^2}}.$$

Reducing by Formula E, integrating the last term by Formula (26), and taking the integral between the limits, $y = 0$, and $y = 2r$, we find, for one-half the volume required,

$$V'' = \frac{5}{2}\pi^2 r^3, \text{ or } 2V'' = 5\pi^2 r^3 = \frac{5}{8}\pi(2r)^2 \times 2\pi r \; \ldots\ldots \; (8)$$

Hence, *the volume is equal to five-eighths the circumscribing cylinder.*

5°. *To find the volume generated by revolving the logarithmic curve about the axis of numbers.*

The equation of the generatrix is $y = lx$. Hence,

$$V = \pi \int (lx)^2 dx \; \ldots\ldots \; (9)$$

Reducing by Formula F, we have,

$$V = \pi[x(lx)^2 - 2(xlx - x)] + C.$$

If the initial plane pass through the point whose abscissa is 1, we have, $C = -2\pi$. Hence,

$$V' = \pi[x(lx)^2 - 2(xlx - x + 1)] \; \ldots\ldots \; (10)$$

PART V.

APPLICATIONS OF THE DIFFERENTIAL AND INTEGRAL CALCULUS TO MECHANICS AND ASTRONOMY.

I. CENTRE OF GRAVITY.

82. In what follows, bodies are supposed to be homogeneous; the weight of any part of a body is therefore proportional to its volume, and consequently the weight of the unit of volume may be taken as the unit of weight. Points, lines, and surfaces are supposed to be material: A *material point* is a body whose length, breadth, and thickness are infinitesimal; a *material line* is a line whose length is finite, and whose breadth and thickness are infinitesimal; a *material surface* is a body whose length and breadth are finite, and whose thickness is infinitesimal; under this supposition a point is an elementary portion of a line, a line is an elementary portion of a surface, and a surface is an elementary portion of a solid.

The weights of the elements of a body are directed toward the centre of the earth, and because the bodies treated of are exceedingly small in comparison with the earth, these weights may be regarded as a system of parallel forces; hence, the weight of a body is equal to the sum of the weights of its elements and is parallel tc them.

The *centre of gravity* of a body is a point through which its weight always passes. This point may be found by the *principle of moments*, which may be enunciated as follows: *the moment of the resultant of any number of forces, with respect to an axis, is equal to the algebraic sum of the moments of the forces, with respect to the same axis.* (*Mechanics*, Art. 35.)

In applying this principle, we assume the following results of demonstrations in mechanics: 1°. The centre of gravity of a straight line is at its middle point; 2°. The centre of gravity of a plane figure is in that plane; 3°. If a plane figure have a line of symmetry, its centre of gravity is on that line; and 4°. If a solid have a plane of symmetry, its centre of gravity is in that plane. (*Mechanics*, Arts. 44, 45, 46.)

To deduce general formulas for finding the centre of gravity of a body, assume a system of co-ordinate axes that are to retain a fixed position with respect to the body, but that change position when the body moves; denote the volume, and consequently the weight, of any element of the body by dv, and the co-ordinates of its centre of gravity by x, y, and z; denote the weight of the body by $v = \int dv$, and the co-ordinates of its centre of gravity by x_1, y_1, and z_1.

If the body be placed in such a position that the plane xy is horizontal, the weights of the elements and of the body are parallel to the axis of z, the moment of the body, with respect to the axis of y, is $x_1 \int dv$, the moment of any element, with respect to the same axis, is xdv

and the algebraic sum of the moments of all the elements is $\int xdv$; hence, from the principle of moments, we have,

$$x_1\int dv = \int xdv\ ; \quad \therefore\ x_1 = \frac{\int xdv}{\int dv} \quad \ldots\ldots (1)$$

In like manner, we have,

$$y_1\int dv = \int ydv\ ; \quad \therefore\ y_1 = \frac{\int ydv}{\int dv} \quad \ldots\ldots (2)$$

If the body be turned about so that the plane yz is horizontal, we have, in like manner,

$$z_1\int dv = \int zdv\ ; \quad \therefore\ z_1 = \frac{\int zdv}{\int dv} \quad \ldots\ldots (3)$$

When the body is in a plane, that plane may be taken as the plane xy, in which case $z_1 = 0$; if the body have an axis of symmetry, that may be taken as the axis of x, in which case $z_1 = 0$, and $y_1 = 0$.

Centre of Gravity of a Circular Arc.

83. Let the radius perpendicular to the chord of the arc be taken as the axis of x: then will z_1, and y_1 be equal to 0. Denote the radius of the circle by r, the chord by c, and the arc by A. The origin being at the centre, the equation of the arc is, $y^2 = r^2 - x^2$; hence,

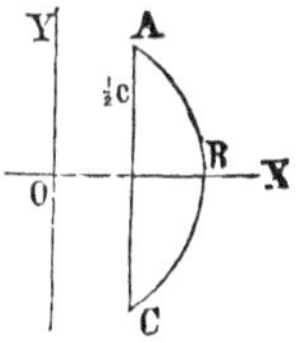

Fig. 16.

$$dv = \sqrt{dx^2 + dy^2} = \sqrt{\frac{y^2}{x^2}dy^2 + dy^2}$$

$$= \frac{rdy}{x} = \frac{rdy}{\sqrt{r^2 - y^2}};$$

Substituting in (1), and integrating between the limits, $y = -\frac{1}{2}c$, and $y = +\frac{1}{2}c$, we have, for the numerator,

$$\int_{-\frac{1}{2}c}^{+\frac{1}{2}c} r dy = rc,$$

and for the denominator,

$$\int_{-\frac{1}{2}c}^{+\frac{1}{2}c} \frac{r dy}{\sqrt{r^2 - y^2}} = r\left[\sin^{-1}\frac{c}{2r} - \sin^{-1}\frac{-c}{2r}\right] = \text{arc } ABC.$$

Hence,

$$x_1 = \frac{rc}{\text{arc } ABC}, \text{ or arc } ABC : c :: r : x_1.$$

That is, *the centre of gravity of the arc of a circle is on the diameter that bisects the chord, and its distance from the centre is a fourth proportional to the arc, its chord, and the radius.*

Centre of Gravity of a Parabolic Area.

84. Let the area be limited by a double ordinate, and denote the extreme abscissa by a. From the equation of the curve, $y^2 = 2px$, we have,

$$y = \sqrt{2p} \cdot x^{\frac{1}{2}}.$$

$$\therefore dv = ydx = \sqrt{2p} \cdot x^{\frac{1}{2}}dx,$$

$$\text{and } xdv = \sqrt{2p} \cdot x^{\frac{3}{2}}dx.$$

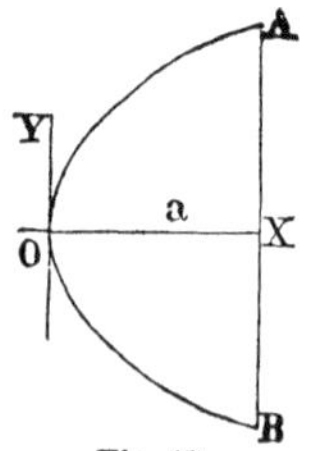

Fig. 17.

Substituting in (1), and integrating between the limits 0 and a, we have,

$$x_1 = \frac{3}{5}a.$$

That is, *the centre of gravity is on the axis, at a distance from the vertex equal to three-fifths the altitude of the segment.*

Centre of Gravity of a Semi-ellipsoid of Revolution.

85. Let the axis of the ellipsoid be taken as the axis of x. Then, if the origin be taken at the centre, the equation of the generating curve is

$$y^2 = \frac{b^2}{a^2}(a^2 - x^2).$$

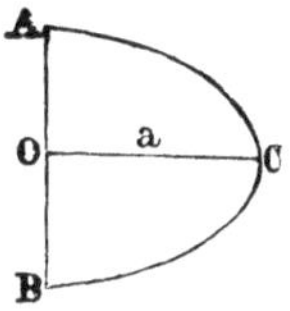

Fig. 18.

In this case, we have,

$$dv = \pi y^2 dx = \pi \frac{b^2}{a^2}(a^2 - x^2)dx,$$

$$\text{and } xdv = \pi \frac{b^2}{a^2}(a^2x - x^3)dx.$$

Substituting in (1), and integrating between the limits $x = 0$, and $x = a$, we have,

$$x_1 = \frac{3}{8}a = \frac{3}{16}2a.$$

That is, *the centre of gravity of a semi-prolate spheroid of revolution is on its axis of revolution, and at a distance from the centre equal to three-sixteenths the major axis of the generating ellipse.*

If we change a to b, and b to a, we find for the semi oblate spheroid,

$$x_1 = \frac{3}{8}b = \frac{3}{16}2b.$$

Centre of Gravity of a Cone.

86. Let the axis of the cone be the axis of x, and the vertex of the cone at the origin; denote the altitude by h, and the radius of the base by r; then will the tangent of the semi-angle of the cone be $\frac{r}{h}$, the equation of the generating line will be $y = \frac{r}{h}x$, and we have,

$$dv = \pi y^2 dx = \pi \frac{r^2}{h^2} x^2 dx, \text{ and } xdv = \pi \frac{r^2}{h^2} x^3 dx.$$

Substituting in (1), and integrating between the limits 0 and h, we have,

$$x_1 = \frac{3}{4} h.$$

Centre of Gravity of a Paraboloid of Revolution.

87. Let the axis of the paraboloid be taken as the axis of x. The equation of the parabola being $y^2 = 2px$, we have,

$$dv = 2\pi pxdx, \text{ and } xdv = 2\pi px^2 dx.$$

Substituting in (1), and integrating from $x = 0$ to $x = a$, we have,

$$x_1 = \frac{2}{3} a.$$

II. Moment of Inertia.

Definitions and Preliminary Principles.

88. The *moment of inertia* of a body with respect to an axis, is equal to the algebraic sum of the products obtained by multiplying the mass of each element of the body by the square of its distance from the axis. If we take the axis through the centre of gravity of the body, and denote the mass of an element by dm, its distance from the axis by x, and the moment of inertia by K, we have,

$$K = \int x^2 dm \quad . \; . \; . \; . \; . \; (1)$$

If we take any parallel axis at a distance d from the assumed axis, and denote the moment of inertia with respect to it by K', we have (*Mechanics*, Art. 123).

$$K' = K + md^2 \quad . \; . \; . \; . \; . \; (2)$$

Moment of Inertia of a Straight Line.

89. Let the axis be taken through the centre of gravity of the line and perpendicular to it. Let AB represent the line, CD the axis, and E any element. Denote the length of the line by $2l$, its mass by m, the distance GE by x, and the length of the element by dx. From the principle of homogeneity, we have,

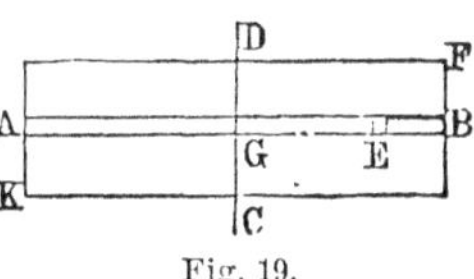

Fig. 19.

$$2l : dx :: m : dm ; \; \therefore \; dm = \frac{m}{2l} dx.$$

Substituting in (1), and integrating from $x = -l$ to $x = +l$, we have,

$$K = m\frac{l^2}{3} ; \; \therefore \; K' = m\left(\frac{l^2}{3} + d^2\right) \quad . \; . \; . \; . \; . \; (3)$$

These formulas are entirely independent of the breadth of AB in the direction of the axis CD; they hold good, therefore, when the filament is replaced by the rectangle KF, the axis being parallel to one of its ends. In this case m is the mass of the rectangle; $2l$, its length; and d, the distance of the axis from the centre of gravity of the rectangle.

Moment of Inertia of a Circle.

90. *First*, let the axis be taken to coincide with one of the diameters. Let ACB represent the circle, AB the axis, and $C'D'$ an element parallel to the axis. Denote OC by r, OE by x, EF by dx, $C'D'$ by $2y$, and the mass of the circle by m. Then, because the circle is homogeneous, we have,

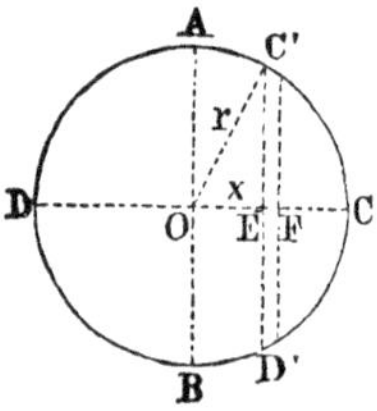

$$\pi r^2 : 2ydx :: m : dm;$$

Fig. 20.

$$\therefore\ dm = \frac{2my}{\pi r^2}dx = \frac{2m\sqrt{r^2 - x^2}}{\pi r^2}dx.$$

Substituting in (1),

$$K = \frac{2m}{\pi r^2}\int (r^2 - x^2)^{\frac{1}{2}}x^2 dx.$$

Reducing by Formulas A and B, integrating by (20), and taking the integral between the limits $x = -r$ and $x = +r$, we find,

$$K = \frac{mr^2}{4};\quad \therefore\ K' = m\left(\frac{r^2}{4} + d^2\right) \ldots\ldots (4)$$

Secondly, let the axis be taken through the centre and perpendicular to the plane of the circle. Let KL be an elementary ring, whose radius is x, and whose breadth is dx; then will its area be $2\pi xdx$, and from the principle, that the masses are proportional to the volumes, we have,

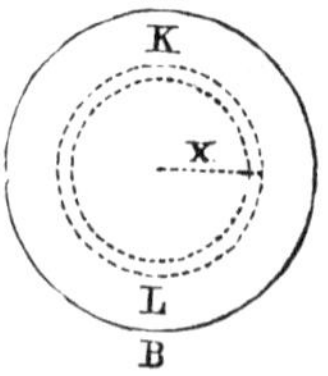

$$\pi r^2 : 2\pi xdx :: m : dm;\ \therefore\ dm = \frac{2mxdx}{r^2}$$

Fig. 21.

Substituting in (1), and integrating from $x = 0$ to $x = r$, we have,

$$K = \frac{mr^2}{2}; \quad \therefore K' = m\left(\frac{r^2}{2} + d^2\right) \; . \; . \; . \; . \; . \; (5)$$

Thirdly, to find the moment of inertia of a circular ring with respect to an axis perpendicular to its plane Let m denote the mass of the ring, r and r' its extreme radii;

Then, $\pi(r^2 - r'^2) : 2\pi x dx :: m : dm$;

$$\therefore dm = \frac{2mx}{r^2 - r'^2}dx.$$

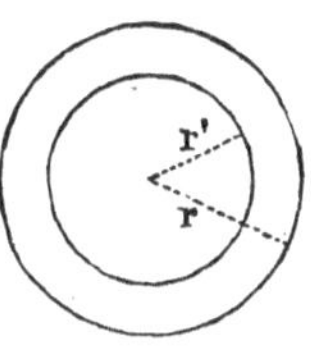

Fig. 22.

Substituting in (1) and integrating from $x = r'$ to $x = r$, we have

$$K = \frac{m(r^2 + r'^2)}{2}; \; \therefore K' = m\left(\frac{r^2 + r'^2}{2} + d^2\right) . \; . \; . \; (6)$$

Formulas (5) and (6) are independent of the thickness of the plate; hence, they will be true whatever that thickness may be. Hence, they hold good for a solid and hollow cylinder, m being taken to represent the mass of the cylinder.

Moment of Inertia of a Cylinder with respect to an Axis perpendicular to the Axis of the Cylinder.

91. Let the axis be taken through the centre of gravity of the cylinder, and let EF be an element perpendicular to the axis of the cylinder. Denote the length of the cylinder by $2l$, its radius by r, its mass by m, the distance of EF from the axis by x, and the thickness by dx. Then, as before, we have,

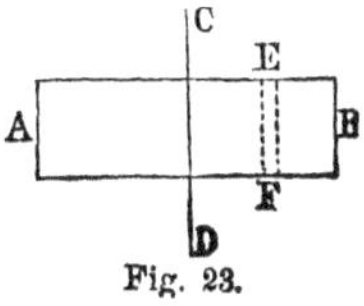

Fig. 23.

$$2l : dx :: m : dm; \quad \therefore dm = \frac{m}{2l}dx.$$

The moment of inertia of this element is given by Formula (4); but this is the differential of the moment of inertia of the entire cylinder; hence, substituting dm for m, and x for d, in (4), we have,

$$dK = \frac{m}{2l}\left(\frac{r^2}{4} + x^2\right)dx.$$

Integrating from $-l$ to $+l$, we have,

$$K = m\left(\frac{r^2}{4} + \frac{l^2}{3}\right); \quad \therefore\ K' = m\left(\frac{r^2}{4} + \frac{l^2}{3} + d^2\right) \quad \ldots\ldots (7)$$

Moment of Inertia of a Sphere.

92. Let the axis pass through the centre of a sphere. Let $C'D'$ be an elementary segment perpendicular to the axis, DC, whose volume is $\pi y^2 dx = \pi(r^2 - x^2)dx$; its mass is found from the proportion,

$$\frac{4}{3}\pi r^3 : \pi(r^2 - x^2)dx :: m : dm;$$

$$\therefore\ dm = \frac{3m}{4r^3}(r^2 - x^2)dx.$$

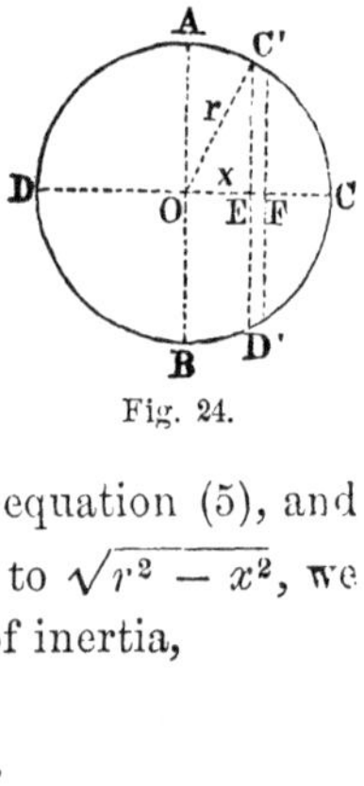

Fig. 24.

Substituting this value of dm for m in equation (5), and making r in that equation equal to y, or to $\sqrt{r^2 - x^2}$, we have, for the differential of the moment of inertia,

$$dK = \frac{3m}{8r^3}(r^2 - x^2)^2 dx.$$

Integrating between the limits $x = -r$, and $x = +r$, we have,

$$K = \frac{2mr^2}{5}; \quad \therefore\ K' = m\left(\frac{2}{5}r^2 + d^2\right).$$

III. Motion of a Material Point.

General Formulas.

93. *Uniform motion* is that in which the moving point passes over equal spaces in equal times; *variable motion* is that in which the moving point passes over unequal spaces in equal times. The *velocity* of a point is its *rate* of motion. The *acceleration* due to a force is the *rate of change* that it produces in the velocity of a point. When a force acts to increase the velocity, the acceleration is *positive*, when it acts to diminish the velocity, the acceleration is *negative*, and *conversely*.

Let us denote the *mass* of a material point by m, its *velocity* at any time t, by v, the *space* moved over at the time t, by s, and the *acceleration* due to the moving force by φ.

When the motion is uniform, *the velocity is constant*, and its measure is the space passed over in any time divided by that time; but we may, in all cases, regard the motion as uniform for an infinitely small time dt; denoting the space described in that time by ds, we have,

$$v = \frac{ds}{dt} \quad . \; . \; . \; . \; . \quad (1)$$

When the velocity varies uniformly *the acceleration is constant*, and we take for its measure the change of velocity in any time divided by that time; but we may, in all cases, regard the velocity as varying uniformly for an infinitely small time dt; denoting the change of velocity in that time by dv, we have,

$$\varphi = \frac{dv}{dt} \quad . \; . \; . \; . \; . \quad (2)$$

In the discussion of motion it is customary to regard the time as the independent variable. Differentiating (1), under that supposition, we find,

$$dv = \frac{d^2 s}{dt};$$

substituting this in (2), we have,

$$\varphi = \frac{d^2 s}{dt^2} \quad \ldots\ldots \quad (3)$$

Equations (1), (2), and (3) are fundamental, and from them we may deduce many laws of motion. If we multiply both members of (3) by m, we have,

$$m\varphi = m\frac{d^2 s}{dt^2}; \text{ or, } F = m\frac{d^2 s}{dt^2} \quad \ldots\ldots \quad (4)$$

In (4), F is the moving force. It will often be convenient to regard the mass of the material point as the unit of mass, in which case, we have, $m = 1$.

Uniformly Varied Motion.

94. *Uniformly varied motion* is that in which the velocity increases, or diminishes, uniformly. In the former case the motion is *uniformly accelerated*, in the latter it is *uniformly retarded*, in both the acceleration is *constant*. Denoting the constant value of φ by g, substituting in (3), multiplying both members by dt, and integrating, we have,

$$\frac{ds}{dt} = gt + C; \text{ or, } v = gt + C \quad \ldots\ldots \quad (5)$$

Multiplying again by dt, and integrating, we find,

$$s = \tfrac{1}{2}gt^2 + Ct + C' \quad \ldots\ldots \quad (6)$$

The constant, C, is what v becomes when $t = 0$; this is called the *initial velocity*, and may be denoted by v'. The constant, C', is what s becomes when $t = 0$; this is called the *initial space*, and may be denoted by s'. Making these substitutions in (5) and (6), we have,

$$v = v' + gt \; . \; . \; . \; . \; . \; (7)$$

$$s = s' + v't + \tfrac{1}{2}gt^2 \; . \; . \; . \; . \; . \; (8)$$

Equations (7) and (8) enable us to discuss all the circumstances of uniformly varied motion (see *Mechanics*, Arts. 103 to 108). If g represent the force of gravity, and v' and s' be each equal to 0, (7) and (8) will express the laws of motion when a body falls from rest under the influence of gravity, regarded as a constant force. Under this supposition they become,

$$v = gt \; . \; . \; . \; . \; . \; (9)$$

$$s = \tfrac{1}{2}gt^2 \; . \; . \; . \; . \; . \; (10)$$

That is, the velocities generated are proportional to the times, and the spaces fallen through are proportional to the squares of the times.

Bodies Falling under the Influence of Gravity, regarded as Variable.

95. In accordance with the Newtonian law, the attraction exerted by the earth on a body at different distances varies inversely as the squares of their distances. Denoting the radius of the earth, supposed a sphere, by r, the force of gravity at the surface by g, any distance from the centre, greater than r, by s, and the force of gravity at that distance by φ, we have,

$$s^2 : r^2 :: g : \varphi; \quad \therefore \varphi = \frac{gr^2}{s^2}.$$

Substituting this value of φ in (3), and at the same time making it negative, because it acts in the direction of s negative, we have,

$$\frac{d^2s}{dt^2} = -\frac{gr^2}{s^2} \quad \ldots\ldots \quad (11)$$

Multiplying by $2ds$, and integrating both members, we have,

$$\frac{ds^2}{dt^2} = \frac{2gr^2}{s} + C; \text{ or, } v^2 = \frac{2gr^2}{s} + C \quad \ldots\ldots \quad (12)$$

If we make $v = 0$, when $s = h$, we have,

$$0 = \frac{2gr^2}{h} + C; \quad \therefore\ C = -\frac{2gr^2}{h},$$

$$\text{and, } v^2 = 2gr^2\left(\frac{1}{s} - \frac{1}{h}\right) \quad \ldots\ldots \quad (13)$$

Equation (13) gives the velocity generated whilst the body is falling from the height h to the height s. If we make $h = \infty$, and $s = r$, (13) becomes,

$$v^2 = 2gr; \quad \therefore\ v = \sqrt{2gr} \quad \ldots\ldots \quad (14)$$

In this equation the resistance of the air is not considered.

If in (14) we make $g = 32.088$ feet, and $r = 20923596$ feet, their equatorial values, we find,

$v = 36644$ feet, or nearly 7 miles per second.

Equation (14) enables us to compute the velocity acquired by a body in falling from an infinite distance to the sun. If we make $g = 890.16$ feet, and $r = 430854.5$ miles, which are their values corresponding to the sun, we find,

$v = 381$ miles per second.

If we make h equal to the distance of Neptune, and s equal to the sun's radius in (13), we find the velocity that a body would acquire in falling from Neptune to the sun, under the influence of the sun's attraction.

To find the time required for a body to fall through any space, substitute $\frac{ds}{dt}$ for v in (13), and solve the results with respect to dt; this gives,

$$dt = -\sqrt{\frac{h}{2gr^2}} \cdot \frac{ds\sqrt{s}}{\sqrt{h-s}} = -\sqrt{\frac{h}{2gr^2}} \cdot \frac{sds}{\sqrt{hs-s^2}} \cdot (15)$$

The negative sign is taken because s decreases as t increases. Reducing (15) by Formula E, and integrating by Formula (26), we find,

$$t = \sqrt{\frac{h}{2gr^2}}\left[(hs-s^2)^{\frac{1}{2}} - \frac{h}{2}\operatorname{versin}^{-1}\frac{2s}{h}\right] + C \ldots\ldots (16)$$

If $t = 0$ when $s = h$, we have,

$$C = \sqrt{\frac{h}{2gr^2}} \times \frac{\pi h}{2}.$$

This, in (16), gives,

$$t = \sqrt{\frac{h}{2gr^2}}\left[(hs-s^2)^{\frac{1}{2}} - \frac{1}{2}h\operatorname{versin}^{-1}\frac{2s}{h} + \tfrac{1}{2}\pi h\right] \ldots (17)$$

Making $s = r$ in (17), we find,

$$t = \sqrt{\frac{h}{2gr^2}}\left[(hr-r^2)^{\frac{1}{2}} - \frac{1}{2}h\operatorname{versin}^{-1}\frac{2r}{h} + \frac{1}{2}\pi h\right] \ldots (18)$$

Which gives the time required for a body to fall from a distance h, to the surface of the sun.

Bodies Falling under the Influence of a Force that varies as the Distance.

96. It will be shown hereafter, that if an opening were made along one of the diameters of the earth, and a body permitted to fall through it, the body would be urged toward the centre by a force varying as the distance from the centre, provided the earth were homogeneous. Assuming that principle, and denoting the force at the surface by g, the radius being r, and the force at the distance, s, from the centre by φ, we have,

$$r : s :: g : \varphi; \therefore \varphi = \frac{g}{r} s.$$

Substituting this value of φ in (3), and at the same time giving it the minus sign, because it acts in the direction of s negative, we have,

$$\frac{d^2 s}{dt^2} = -\frac{gs}{r}.$$

Multiplying by $2ds$, and integrating,

$$\frac{ds^2}{dt^2} = -\frac{gs^2}{r} + C; \text{ or, } v^2 = C - \frac{g}{r} s^2.$$

Making $v = 0$, when $s = r$, we have, $C = \frac{g}{r} r^2$; hence,

$$\frac{ds^2}{dt^2} = \frac{g}{r}(r^2 - s^2) \quad \ldots\ldots \quad (19)$$

If $s = 0$, we find $v = \sqrt{gr}$, which is the maximum velocity. It is equal to the entire velocity generated by a body falling from an infinite distance to the surface of the earth divided by $\sqrt{2}$. If the body pass the centre, s becomes negative, and we find the same values for v at equal distances from the centre, whether s be positive or negative. When s becomes equal to $-r$, v reduces to 0, and the body then falls toward the centre again, and so on continually.

Let us suppose $A'C'$ to be the diameter along which the body oscillates, and at the time the body starts from A' let a second body start from the same point and move around the semi-circumference, $A'MC'$, with a constant velocity equal to $\sqrt{gr}$. At any point, M, let this velocity be resolved into two components, MQ and MN, the former perpendicular and the latter parallel to $A'C'$. The latter component will be equal to $\sqrt{gr}$, multiplied by $\cos TMN$, or its equal, $\cos B'MH'$; but $\cos B'MH' = \frac{H'M}{B'M}$; denoting $B'H'$ by s, whence $H'M = \sqrt{r^2 - s^2}$, we have, $\cos B'MH' = \frac{\sqrt{r^2 - s^2}}{r}$; hence, $MN = \sqrt{\frac{g}{r}} \cdot (r^2 - s^2)^{\frac{1}{2}}$; but this is equal to the velocity of the oscillating point when at H'. Hence, the velocity of the vibrating point is everywhere equal to the parallel component of the velocity of the revolving point; they will, therefore, come together at the points A' and C', and the position of the vibrating point will always be found by projecting the corresponding position of the revolving point on the path of the former. To find the time for a complete vibration from A' to C', solve (19) with respect to dt, whence,

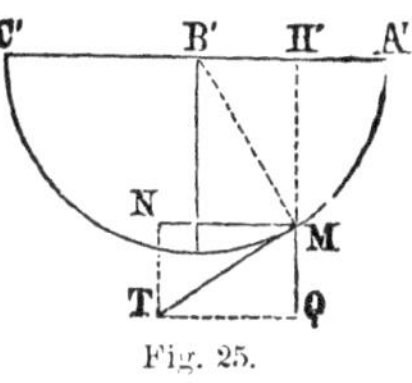

Fig. 25.

$$dt = -\sqrt{\frac{r}{g}} \cdot \frac{ds}{\sqrt{r^2 - s^2}} \quad \ldots\ldots \quad (20)$$

The negative sign is used because t is a decreasing function of s. Integrating (20) between the limits $s = r$ and $s = -r$, we find,

$$t'' = \pi\sqrt{\frac{r}{g}} \quad . \quad . \quad . \quad . \quad . \quad (21)$$

This, we shall see hereafter, is the time of vibration of a simple pendulum whose length is r.

The time may be found otherwise, as follows: The space passed over by the revolving point is πr; dividing this by the velocity, $\sqrt{gr}$, we have,

$$t = \pi\sqrt{\frac{r}{g}}.$$

The species of vibratory motion just discussed is sometimes called harmonic, being the same as that of a point of a vibrating chord or spring.

Vibration of a Particle of an Elastic Medium.

97. It is assumed that if a particle of an elastic medium be slightly disturbed from its place of rest, and then abandoned, it will be urged back by a force that varies *directly* as the distance of the particle from its position of equilibrium; on reaching this position, the particle, by virtue of its inertia, will pass to the other side, again to be urged back, and so on.

Let us denote the displacement, at any time t, by s, and the acceleration due to the restoring force by φ; then, from the law of force, we have, $\varphi = n^2 s$, in which n is constant for the same medium, under the same circumstances of density, pressure, etc. Substituting for φ its value from Equation (3), and prefixing the negative sign, because it acts in a direction contrary to that in which s is estimated, we have,

$$-\frac{d^2s}{dt^2} = n^2s \quad \ldots\ldots \quad (22)$$

Multiplying by $2ds$, and integrating, we have,

$$-\frac{ds^2}{dt^2} = n^2s^2 + C = -v^2 \quad \ldots\ldots \quad (23)$$

The velocity is 0 when the particle is at the greatest distance from the position of equilibrium; denoting this value of s by a, we have.

$$n^2a^2 + C = 0; \quad \therefore \ C = -n^2a^2,$$

which, in (23), gives,

$$\frac{ds^2}{dt^2} = n^2(a^2 - s^2)\text{; or, } ndt = \frac{ds}{\sqrt{a^2 - s^2}} \quad \ldots\ldots \quad \mathbf{(24)}$$

Integrating (24), we have.

$$nt + C' = \sin^{-1}\frac{s}{a} \quad \ldots\ldots \quad (25)$$

Taking the sines of both members, and reducing, we have,

$$s = a\sin(nt + C') \quad \ldots\ldots \quad (26)$$

If we make $t = 0$, when $s = 0$, we have $C' = 0$, and Equation (26) becomes,

$$s = a\sin(nt) \quad \ldots\ldots \quad (27)$$

If (25) be taken between the limits $-a$ and $+a$, we find for the time of a single vibration, denoted by $\frac{1}{2}T$,

$$\tfrac{1}{2}T = \frac{\pi}{n}; \quad \therefore \ n = \frac{2\pi}{T}.$$

Substituting this in (27), we have, finally,

$$s = a \sin\left(\frac{2\pi}{T} t\right) \quad . \quad . \quad . \quad . \quad . \quad (28)$$

in which T is the time of a double vibration.

Solving (24), we have,

$$\frac{ds}{dt} = v = n\sqrt{a^2 - s^2} \quad . \quad . \quad . \quad . \quad . \quad (29)$$

Substituting for s, in (29), its value from (28), we find,

$$v = n\sqrt{a^2 - a^2 \sin^2\left(\frac{2\pi}{T} t\right)} = na\sqrt{1 - \sin^2\left(\frac{2\pi}{T} t\right)}$$

Whence, we have,

$$v = na \cos\left(\frac{2\pi}{T} t\right) \quad . \quad . \quad . \quad . \quad . \quad (30)$$

This equation is used in discussing the laws of light, and in many other cases.

Curvilinear Motion of a Point.

98. A point cannot move in a curve except under the action of an incessant force, whose direction is inclined to the direction of the motion. This force is called the *deflecting force,* and can be resolved into two components, one in the direction of the motion, and the other at right angles to it. The former acts simply to increase or diminish the velocity, and is called the *tangential force;* the latter acts to turn the point from its rectilinear direction, and being directed toward the centre of curvature is called the *centripetal force.* The resistance offered by the point to the centripetal force, in consequence of its inertia, is equal and directly opposed to the centripetal force, and is called the *centrifugal force.*

To find expressions for the tangential and centripetal forces, let the acceleration due to the deflecting force in the direction of the axis of x, at any time t, be denoted by X, and the acceleration in the direction of the axis of y by Y. Let these be resolved into components acting tangentially and normally. The algebraic sum of the tangential components is the tangential acceleration denoted by T, and the algebraic sum of the normal components is the centripetal acceleration denoted by N. Assuming the notation of the figure, we have,

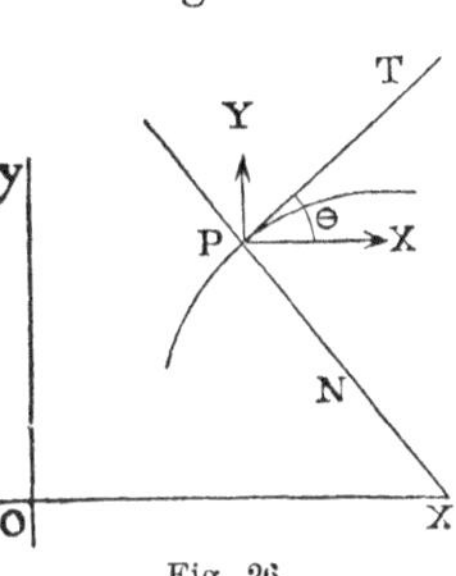

Fig. 26.

$$T = X\cos\theta + Y\sin\theta;$$

$$N = X\sin\theta - Y\cos\theta.$$

But from Articles (93) and (5), we have,

$$X = \frac{d^2x}{dt^2}; \quad Y = \frac{d^2y}{dt^2};$$

and,

$$\cos\theta = \frac{dx}{ds}; \quad \sin\theta = \frac{dy}{ds}.$$

Substituting in the preceding equations, and reducing on the supposition that t is the independent variable, and by means of Formula (3), Art. (38), we have,

$$T = \frac{d^2x}{dt^2} \cdot \frac{dx}{ds} + \frac{d^2y}{dt^2} \cdot \frac{dy}{ds}$$

$$= \frac{d(dx^2 + dy^2)}{dt^2 \,.\, 2ds} = \frac{d(ds^2)}{dt^2 2ds} = \frac{d^2s}{dt^2} \quad . \; . \; . \; . \; . \quad (31)$$

$$N = \frac{d^2xdy - d^2ydx}{dt^2 \cdot ds}$$

$$= -\frac{ds^2}{dt^2} \cdot \frac{d^2ydx - d^2xdy}{ds^3} = -\frac{v^2}{R}$$

But the centripetal acceleration is equal and directly opposed to the centrifugal acceleration. Denoting the latter by f, we have,

$$f = \frac{v^2}{R} \quad \ldots\ldots \quad (32)$$

From (31) we see that the tangential force is independent of the centripetal force, and from (32) we see that the *acceleration due to the centrifugal force at any point of the trajectory, is equal to the square of the velocity divided by the radius of curvature at that point.*

Velocity of a Point rolling down a Curve in a Vertical Plane.

99. Let a point roll down a curve, situated in a vertical plane, under the influence of gravity regarded as constant. Let the origin of co-ordinates be taken at the starting point, and let distances downward be positive. At any point of the curve, whose ordinate is y, the force of gravity being denoted by g, we have, for the tangential component of gravity, $g\sin\delta$, or, $g\frac{dy}{ds}$; placing this equal to its value, equation (31), we have,

$$g\frac{dy}{ds} = \frac{d^2s}{dt^2}, \text{ or } gdy = \frac{ds \cdot d^2s}{dt^2}.$$

Integrating and reducing, remembering that the constant is 0 under the particular hypothesis, we have,

$$gy = \frac{1}{2}\frac{ds^2}{dt^2}; \text{ or, } v^2 = 2gy. \quad \therefore v = \sqrt{2gy} \quad \ldots\ldots \quad (1)$$

The second member of (1) is the velocity due to the height y. Hence, the velocity generated by a body rolling down a curve, gravity being constant, is equal to that generated by falling freely through the same vertical height.

This principle is true so long as g is constant. But g may be regarded as constant from element to element, no matter what may be the law of variation. Hence, if a body fall toward the sun, or earth, on a spiral line, the ultimate velocity will be the same as though it had fallen on a right line toward the centre of the attracting body. The direction of the motion, however, is not the same, for in the former case it is tangential to the trajectory pursued, and in the latter case it is normal to the attracting body.

The Simple Pendulum.

100. A simple pendulum is a material point suspended from a horizontal axis, by a line without weight, and free to vibrate about that axis.

Let ABC be the arc through which the vibration takes place, and denote its radius by l. The angle CDA is the *amplitude* of vibration; half this angle, ADB, denoted by a, is the *angle of deviation;* and l is *the length of the pendulum.* If the point start from rest, at A, it will, on reaching any point, H, of its path, have a velocity, v, due to the height EK, denoted by y. Hence,

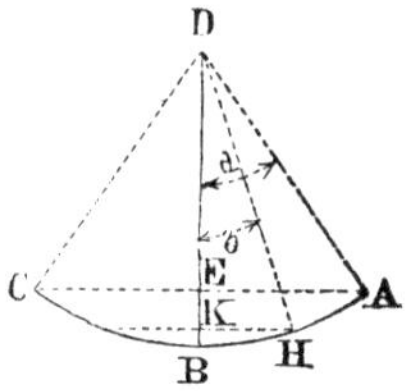

Fig. 27.

$$v = \sqrt{2gy}.$$

If we denote the angle HDB by θ, we have $DK = l\cos\theta$; we also have $DE = l\cos\alpha$; and since y is equal to $DK - DE$, we have,

$$y = l(\cos\theta - \cos\alpha),$$

which, being substituted in the preceding formula, gives,

$$v = \sqrt{2gl(\cos\theta - \cos\alpha)}.$$

Again, denoting the angular velocity at the time t by φ, and remembering that $\varphi = \frac{d\theta}{dt}$, we have, for the velocity of the pendulum,

$$v = l \times \varphi = l\frac{d\theta}{dt}.$$

Equating the two values of v, and reducing, we have,

$$dt = \sqrt{\frac{l}{2g}} \cdot \frac{d\theta}{\sqrt{\cos\theta - \cos\alpha}}.$$

Developing $\cos\theta$ and $\cos\alpha$ by McLaurin's Formula, we have,

$$\cos\theta = 1 - \frac{\theta^2}{2} + \frac{\theta^4}{24} - \text{etc.}$$

$$\cos\alpha = 1 - \frac{\alpha^2}{2} + \frac{\alpha^4}{24} - \text{etc.}$$

If we suppose the amplitude to be small, we may neglect all the terms after the second; doing so, and substituting in (31), we have,

$$dt = \sqrt{\frac{l}{g}} \cdot \frac{d\theta}{\sqrt{\alpha^2 - \theta^2}}.$$

Integrating between the limits $\theta = -\alpha$, and $\theta = +\alpha$, we find,

$$t = \pi\sqrt{\frac{l}{g}}.$$

which is the formula for the time of vibration of a simple pendulum.

Attraction of Homogeneous Spheres.

101. The Newtonian Law of universal gravitation may be expressed as follows, viz.: every particle of matter attracts every other particle, with a force that varies *directly* as the mass of the attracting particle, and *inversely* as the square of the distance between the particles. To apply this law to the case of homogeneous spheres, let us first consider the action of a spherical shell, of infinitesimal thickness, on a material point within it.

Let D be the material point, APB a great circle through it, and let the diameter ADB be taken as the axis of x. If the circle be revolved about AB, its circumference will generate a spherical shell, and any element of the circumference, as P, will generate an elementary zone whose altitude is dx. For any point of this zone, as P, there is another point, P', symmetrical with it, and the resultant action of these points on D is directed along AB, and is equal to the sum of the forces into the cosine of the angle BDP. Hence, the resultant attraction of the entire zone is directed along AB, and is equal to the sum of the attractions of all its particles into the cosine of the angle BDP. To find an expression for this resultant, denote CP by r, CD by a, DP by z, and CE by x. Because the shell is infinitesimal in thickness and homogeneous, the mass of the zone, is to the mass of the shell, as dx, the altitude of the zone, is to $2r$, the altitude of the shell. Hence, if the mass of the shell be

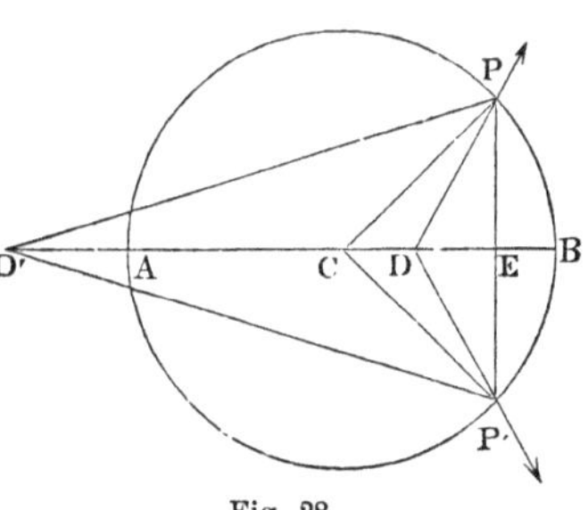

Fig. 28.

denoted by m, the mass of the zone will be denoted by $\frac{mdx}{2r}$. If the force exerted by the unit of mass at the unit of distance be taken as the unit of force, the attraction exerted by all the particles of the zone on the point D will be equal to $\frac{mdx}{2r} \cdot \frac{1}{z^2}$, and the resultant action on D, in the direction of AB, denoted by df, will be given by the equation,

$$df = \frac{mdx}{2r} \cdot \frac{1}{z^2} \cdot \cos EDP \quad . \; . \; . \; . \; . \quad (1)$$

From the triangle, PCD, we have,

$$\overline{DP}^2 = z^2 = r^2 + a^2 - 2ax = r^2 - a^2 - 2a(x - a),$$

and from the triangle, EDP, we have,

$$\cos EDP = \frac{x - a}{z}.$$

Substituting in (1), remembering that dx equals $d(x - a)$, we have,

$$df = \frac{m}{2r} \cdot \frac{(x - a)\, d(x - a)}{[r^2 - a^2 - 2a(x - a)]^{\frac{3}{2}}} \quad . \; . \; . \; . \; . \quad (2)$$

Regarding $(x - a)$ as a single variable, reducing by Formula A, and integrating by Formula (29), we find,

$$F = \frac{m}{2r} \left\{ - \frac{x - a}{a[r^2 - a^2 - 2a(x - a)]^{\frac{1}{2}}} + \frac{r^2 - a^2}{a^2[r^2 - a^2 - 2a(x - a)]^{\frac{1}{2}}} \right\}.$$

Taking the integral from $x = -r$ to $x = +r$, we have,

$$f'' = 0$$

Hence, the effect of the attraction of the shell on any point within it is null. If a sphere be described about C as a centre with a radius equal to a, we may call that part which lies between it and the surface of the given sphere the *exterior shell*, and the sphere itself may be called the *nucleus*. From what precedes, we infer that any point within a homogeneous sphere is acted on by the sphere precisely as though the exterior shell did not exist. Hence, a point at the centre of a sphere is not affected by the attraction of the sphere.

If the point D' be taken without the shell, we have,

$$\overline{D'P}^2 = z^2 = r^2 + a^2 + 2ax = r^2 - a^2 + 2a(x + a),$$

and from the triangle, $ED'P$, we have,

$$\cos ED'P = \frac{x + a}{z}.$$

Substituting in (1), we have,

$$df = \frac{m}{2r} \frac{(x + a)d(x + a)}{[r^2 - a^2 + 2a(x + a)]^{\frac{3}{2}}} \cdot \ldots\ldots (3)$$

Reducing and integrating as before, we find,

$$f = \frac{m}{2r} \left\{ \frac{x + a}{a[r^2 - a^2 + 2a(x + a)]^{\frac{1}{2}}} + \frac{r^2 - a^2}{a^2[r^2 - a^2 + 2a(x + a)]^{\frac{1}{2}}} \right\} + C.$$

Taking the integral between the limits $x = -r$ and $x = +r$, we have, when $a > r$

$$f' = \frac{m}{a^2} \cdot \ldots\ldots (4)$$

But m is the mass of the shell, and a is the distance of D from the centre; hence, the shell attracts a particle without it as though its entire mass were concentrated at its centre.

A homogeneous sphere may be regarded as made up of spherical shells; hence, a sphere attracts any point without it, as though the mass of the sphere were concentrated at its centre. The same is true of a sphere made up of homogeneous strata, which vary in density in passing from the surface to the centre. It is to be inferred that two homogeneous spheres, or two spheres made up of homogeneous strata, attract each other as though both were concentrated at their centres.

If an opening were made from the surface to the centre of the earth, supposed homogeneous, and a body were to move along it under the earth's attraction, it would everywhere be urged on by a force varying *directly* as the distance of the body from the centre. For, denote the distance of the body from the centre at any instant by x. The body will only be acted on by the nucleus whose radius is x. If we take r to represent the radius of the earth, and remember that the masses are proportional to these volumes, we have,

$$M : m :: \frac{4}{3}\pi r^3 : \frac{4}{3}\pi x^3 :: r^3 : x^3.$$

Denoting the force of attraction at the surface by g, and the force of attraction at the point whose distance from the centre is x by f, we have, from the Newtonian law,

$$g : f :: \frac{r^3}{r^2} : \frac{x^3}{x^2} :: r : x; \quad \therefore f = \frac{g}{r}x.$$

Hence, the proposition is proved. (See Art. 96.)

Orbital Motion.

102. If a moving point, P, be continually acted on by a deflecting force directed toward a fixed centre, it will describe a line or path, called an *orbit*. If the *moving* point be undisturbed by the action of any other force, the orbit will lie in a plane passing through the fixed centre and an element of the curve. Let this plane be taken as the co-ordinate plane, let the fixed point be the origin, and let the orbit be represented by APB.

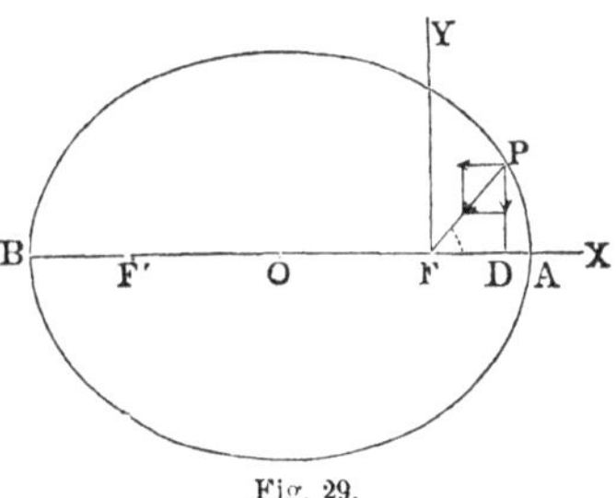

Fig. 29.

Denote the acceleration due to the deflecting force at any time t by f, its inclination to the axis by φ, and its components in the directions of the co-ordinate axes by X and Y.

From the figure, we have,

$$X = -f\cos\varphi, \text{ and } Y = -f\sin\varphi \quad . \; . \; . \; . \; . \quad (1)$$

If we regard t as the independent variable, dx and dy will both be variable, and from Equation (3), Art 93, we have,

$$X = \frac{d^2x}{dt^2}, \text{ and } Y = \frac{d^2y}{dt^2}.$$

If we denote the co-ordinates of P by x and y, and its radius vector FP by r, we have, from the figure,

$$\cos\varphi = \frac{x}{r}, \; \sin\varphi = \frac{y}{r};$$

and $$x^2 + y^2 = r^2; \quad \therefore \; xdx + ydy = rdr.$$

Substituting in (1), we have,

$$\frac{d^2x}{dt^2} = -f\frac{x}{r}, \text{ and } \frac{d^2y}{dt^2} = -f\frac{y}{r} \quad \ldots\ldots \quad (2)$$

Multiplying the first of Equations (2) by y, the second by x, and subtracting the former from the latter, we have,

$$\frac{xd^2y}{dt^2} - \frac{yd^2x}{dt^2} = 0, \quad \text{or} \quad \frac{d\,(xdy - ydx)}{dt} = 0 \quad \ldots\ldots \quad (3)$$

Multiplying the first by dx, the second by dy, adding the resulting equations and reducing, we have,

$$\frac{dxd^2x}{dt^2} + \frac{dyd^2y}{dt^2} = -f\frac{xdx + ydy}{r} = -fdr \quad \ldots\ldots \quad (4)$$

Integrating (3) and (4), Art. 77, we have,

$$x\frac{dy}{dt} - y\frac{dx}{dt} = C \quad \ldots\ldots \quad (5)$$

and

$$\frac{1}{2}\frac{dx^2 + dy^2}{dt^2} + C' = -\int fdr;$$

or,

$$\frac{ds^2}{dt^2} + 2\int fdr = C'' \quad \ldots\ldots \quad (6)$$

Equations (5) and (6) make known the circumstances of motion when the value of f is given. It is found convenient to transform them to a system of polar co-ordinates, whose pole is F, and whose initial line is FX. The formulas for transformation are

$$x = r\cos\varphi, \text{ and } y = r\sin\varphi \quad \ldots\ldots \quad (7)$$

Hence, by differentiating, we have,

$$\frac{dy}{dt} = \frac{dr}{dt}\sin\varphi + r\cos\varphi\frac{d\varphi}{dt} \quad \ldots\ldots \quad (8)$$

and

$$\frac{dx}{dt} = \frac{dr}{dt}\cos\varphi - r\sin\varphi\frac{d\varphi}{dt} \quad \ldots\ldots \quad (9)$$

We also, have, Art. (53),

$$ds^2 = r^2 d\varphi^2 + dr^2 \; . \; . \; . \; . \; . \; . \; (10)$$

Substituting in (5), and reducing by the relations,

$$x\sin\varphi - y\cos\varphi = 0, \text{ and } x\cos\varphi + y\sin\varphi = \frac{x^2 + y^2}{r} = r,$$

we have,

$$\frac{r^2 d\varphi}{dt} = C \; . \; . \; . \; . \; . \; (11)$$

From equation (6), we have,

$$\frac{dr^2 + r^2 d\varphi^2}{dt^2} + 2\int f dr = C'' \; . \; . \; . \; . \; . \; (12)$$

But, from (11), we have,

$$dt = \frac{r^2 d\varphi}{C}, \text{ or, } dt^2 = \frac{r^4 d\varphi^2}{C^2},$$

and this, in (12), gives,

$$C^2 \left\{ \frac{dr^2}{r^4 d\varphi^2} + \frac{1}{r^2} \right\} + 2\int f dr = C'' \; . \; . \; . \; . \; . \; (13)$$

Equations (11) and (13) are the equations required. Multiplying (11) by dt, and integrating, we have.

$$\int r^2 d\varphi = Ct + C''' \; . \; . \; . \; . \; . \; (14)$$

But, by Art. 53, $r^2 d\varphi$ is twice the elementary area swept over by the radius vector; hence, the first member of (14) is an expression for twice the area swept over by the radius vector up to the time t. If we suppose the area to be reckoned from the initial line FX, and at the same time suppose $t = 0$, we have $C''' = 0$. Substituting this in (14), and in the resulting equation making $t = 1$, we find C equal to twice the area swept over in the first unit of time; denoting the area described in the unit of time by A, we have, $C = 2A$, and consequently,

$$\int r^2 d\varphi = 2At \quad \ldots\ldots \quad (15)$$

From Equation (15) we infer that the areas described by the radius vector are proportional to the times of description, and this without reference to the nature of the deflecting force.

To find the equation of the orbit, let us assume as a particular case, the Newtonian law of universal gravitation. Denoting the force exerted by the central body at a unit's distance by k, we have, for the attraction at the distance r,

$$f = \frac{k}{r^2} \quad \ldots\ldots \quad (16)$$

Making $r = \frac{1}{u}$, whence $dr = -\frac{du}{u^2}$, and substituting for C and f their values in (13), we have,

$$4A^2 \left\{ \frac{du^2}{d\varphi^2} + u^2 \right\} - 2ku = C^{\text{iv}} \quad \ldots\ldots \quad (17)$$

or,

$$\frac{du^2}{d\varphi^2} + u^2 = \frac{2k}{4A^2}u + C^{\text{v}} \quad \ldots\ldots \quad (18)$$

To find the value of C^{v}, let $\frac{du}{d\varphi} = 0$, when $r = r'$, or $u = u'$. But when $\frac{du}{d\varphi} = 0$, we also have $\frac{dr}{d\varphi} = 0$, that is, the radius vector is perpendicular to the arc. If we denote the corresponding velocity by v', we have,

$$A = \frac{v'r'}{2}, \text{ or, } 4A^2 = v'^2 r'^2;$$

and, consequently,

$$\frac{2k}{4A^2}u = \frac{2k}{v'^2 r'^3}.$$

Making these substitutions in (18), we find,

$$C^{\text{V}} = \frac{1}{r'^2} - \frac{2k}{v'^2 r'^3}.$$

Denoting this value of C^{V} by S, and solving Equation (18), we have,

$$\frac{du^2}{d\varphi^2} = S - \left(u^2 - \frac{2k}{4A^2}u\right) = S + \frac{k^2}{16A^4} - \left(u - \frac{k}{4A^2}\right)^2$$
$$= p^2 - (u - q)^2.$$

Solving with reference to φ, and taking the negative sign of the radical, we have,

$$d\varphi = -\frac{du}{\sqrt{p^2 - (u-q)^2}} = -\frac{d(u - q)}{\sqrt{p^2 - (u-q)^2}} \quad \ldots . \ (19)$$

Integrating, we have,

$$\varphi - \alpha = \cos^{-1}\frac{u - q}{p} \quad \ldots \ldots \ (20)$$

In which $-\alpha$ is an arbitrary constant. If we suppose $\varphi = 0$, when $v = v'$, $r = r'$, and $u = u'$, the corresponding value of α is the angular distance from the initial line to the radius vector whose direction is perpendicular to the element of the curve.

Taking the cosines of both members of (20), we have,

$$\frac{u-q}{p} = \cos(\varphi - \alpha), \text{ or, } u = q + p\cos(\varphi - \alpha)$$

and finally replacing u by its value, and solving, we have,

$$r = \frac{1}{q + p\cos(\varphi - \alpha)} = \frac{4A^2}{k + \sqrt{16SA^4 + k^2} \times \cos(\varphi - \alpha)} \quad \ldots\ldots (21)$$

Equation (21) is the polar equation of a conic section. Hence, the orbit of a particle about a central attracting body is one of the conic sections. If the orbit is a closed curve, it must be an ellipse. This corresponds to the case of a planet revolving about the sun, or to that of a satellite revolving about its primary.

Law of Force.

103. If the orbit of a revolving particle be an ellipse, it must be attracted to the focal point by a force that varies inversely as the square of the distance.

The polar equation of an ellipse is

$$r = \frac{a(1 - e^2)}{1 + e\cos(\varphi - \alpha)} \quad \ldots\ldots (1)$$

in which a is the semi-transverse axis, e is the eccentricity, α is the angular distance from the fixed line to the radius vector drawn to the nearest vertex, or the longitude of the perihelion in the case of a planetary orbit. The angle, $\varphi - \alpha$, is the angle that is called in astronomy the true anomaly.

Putting $r = \frac{1}{u}$, in (1), we get,

$$u = \frac{1 + e\cos(\varphi - \alpha)}{a(1 - e^2)} \quad \ldots\ldots (2)$$

Differentiating twice, with respect to φ, we find,

$$\frac{du}{d\varphi} = -\frac{e\sin(\varphi - \alpha)}{a(1 - e^2)} \quad \ldots\ldots \quad (3)$$

and,

$$\frac{d^2u}{d\varphi^2} = -\frac{e\cos(\varphi - \alpha)}{a(1 - e^2)} \quad \ldots\ldots \quad (4)$$

Adding (2) and (4), we have,

$$\frac{d^2u}{d\varphi^2} + u = \frac{1}{a(1 - e^2)} \quad \ldots\ldots \quad (5)$$

Resuming equation (13) of the last article, replacing r by its value, $\frac{1}{u}$, and dr by its value, $-\frac{du}{u^2}$, differentiating and reducing, we have,

$$C^2 \left\{ \frac{d^2u}{d\varphi^2} + u \right\} - \frac{f}{u^2} = 0 \quad \ldots\ldots \quad (6)$$

Combining (5) and (6), and solving with respect to f, we have,

$$f = \frac{C^2u^2}{a(1 - e^2)}, \text{ or } f = \frac{C^2}{a(1 - e^2)} \cdot \frac{1}{r^2} \quad \ldots\ldots \quad (7)$$

Hence, f varies inversely as the square of r, which was to be shown.

Note on the Methods of the Calculus.

104. All the rules and principles of the Calculus, in the present treatise, have been deduced in accordance with *the method of infinitesimals*, as explained in Articles 6 and 7. They might also have been deduced by *the method of limits*. It remains to be shown that the results obtained by these two methods are always *identically* the same.

To explain the *method of limits*, let us denote any function of x by y; that is, let us assume the equation,

$$y = f(x) \quad . \quad . \quad . \quad . \quad . \quad (1)$$

If we increase x by a variable increment h, and denote the corresponding value of y by y', we have,

$$y' = f(x + h) \quad . \quad . \quad . \quad . \quad . \quad (2)$$

It is shown in Courtenay's Calculus, Article 4, that so long as x retains its general value, the new state of the function can be expressed by the formula,

$$y' = f(x + h) = f(x) + Ah + Bh^2 + Ch^3 + \text{etc.} \,.\,.\, (3)$$

in which A, B, C, etc., depend on x, but are independent of h. If we subtract (1) from (3), member from member, we have,

$$y' - y = Ah + Bh^2 + (\text{etc.})\, h^3 \quad . \quad . \quad . \quad . \quad . \quad (4)$$

Dividing both members of (4) by h, we have,

$$\frac{y' - y}{h} = A + Bh + (\text{etc.})\, h^2 \quad . \quad . \quad . \quad . \quad . \quad (5)$$

The first member of (5) is a symbol to express the ratio of *the increment of the variable, to the corresponding increment of the function*, and the second member is the value of that ratio.

It is shown in Algebra, that in an expression like the second member of equation (5), it is always possible to give to h a value small enough to make the first term numerically greater than the algebraic sum of all the others. If we assign such a value to h, and then suppose h to go on diminishing, the second member will continually approach A, and when h becomes 0, the second member will reduce to A. Hence, A is a quantity toward which the ratio, $\frac{y'-y}{h}$, approaches as h is diminished, but beyond which it cannot pass; it is therefore the *limit* of that ratio. This limit is *the differential coefficient of y*, and, as may easily be seen, it is entirely independent of dx. The product of this by the differential of the variable is *the differential of y*.

We may, therefore, enunciate *the method of limits* as follows: viz., *Give to the independent variable a variable increment, and find the corresponding state of the function; from this subtract the primitive state, and divide the difference by the variable increment; then pass to the limit of the quotient by making the increment of the variable equal to 0; the result is the differential coefficient of the function; if this be multiplied by the differential of the variable, the product is the differential of the function.*

In the case assumed, we have, by this method,

$$\frac{dy}{dx} = A\,; \quad \therefore\ dy = Adx \quad . \ . \ . \ . \ . \quad (6)$$

If we make $h = dx$, in Equation (4), dx being infinitely small, the first member will be the differential of y, and all the terms of the second member after the first may be neglected. Hence, by the method of infinitesimals, we have,

$$dy = Adx\,; \quad \therefore\ \frac{dy}{dx} = A.$$

This result is the same as that obtained by the method of limits. But, by hypothesis, y represents any function of x; hence, in all cases, the differential coefficient is identically the same whether found by the method of limits or by the method of infinitesimals.

Every function, regarded as a *primitive*, is connected with some other function, regarded as a *derivative*, by the law of differentiation. This derivative is the differential coefficient of the primitive. The object of the differential calculus is to find the derivative from its primitive; the object of the integral calculus is to find the primitive from its derivative; every application of the calculus depends on one of these processes, or on some discussion growing out of one, or the other. Now, because the primitive and the derivative are independent of the differentials of both function and variable, the relation between them is, of necessity, independent of the methods employed in establishing that relation. But it has been shown that the same relation is found between these functions, whether we employ the method of infinitesimals or the method of limits. Hence, these methods, and the results obtained by them, are in all cases logically identical.

MATHEMATICS.

DAVIES'S COMPLETE SERIES.

ARITHMETIC.

Davies' Primary Arithmetic.
Davies' Intellectual Arithmetic.
Davies' Elements of Written Arithmetic.
Davies' Practical Arithmetic.
Davies' University Arithmetic.

TWO-BOOK SERIES.

First Book in Arithmetic, Primary and Mental.
Complete Arithmetic.

ALGEBRA.

Davies' New Elementary Algebra.
Davies' University Algebra.
Davies' New Bourdon's Algebra.

GEOMETRY.

Davies' Elementary Geometry and Trigonometry.
Davies' Legendre's Geometry.
Davies' Analytical Geometry and Calculus.
Davies' Descriptive Geometry.
Davies' New Calculus.

MENSURATION.

Davies' Practical Mathematics and Mensuration.
Davies' Elements of Surveying.
Davies' Shades, Shadows, and Perspective.

MATHEMATICAL SCIENCE.

Davies' Grammar of Arithmetic.
Davies' Outlines of Mathematical Science.
Davies' Nature and Utility of Mathematics.
Davies' Metric System.
Davies & Peck's Dictionary of Mathematics.

DAVIES' SERIES—*Continued.*

THE NEW SURVEYING.

Van Amringe's Davies' Surveying.

By Charles Davies, LL.D., author of a Full Course of Mathematics. Revised by J Howard Van Amringe, A.M., Ph.D., Professor of Mathematics in Columbia College 535 pages. 8vo. Full sheep.

Davies' Surveying originally appeared as a text-book for the use of the United States Military Academy at West Point. It proved acceptable to a much wider field, and underwent changes and improvements, until the author's final revision, and has remained the standard work on the subject for many years.

In the present edition, 1883, while the admirable features which have hitherto commended the work so highly to institutions of learning and to practical surveyors have been retained, some of the topics have been abridged in treatment, and some enlarged. Others have been added, and the whole has been arranged in the order of progressive development. A change which must prove particularly acceptable is the transformation of the article on mining-surveying into a complete treatise, in which the location of claims on the surface, the latest and best methods of underground traversing, &c., the calculation of ore-reserves, and all that pertains to the work of the mining-surveyor, are fully explained and illustrated by practical examples. Immediately on the publication of this edition it was loudly welcomed in all quarters. A letter received as we write, from Prof. R. C. Carpenter, of the Michigan State Agricultural College, says: "I am delighted with it. I do not know of a more complete work on the subject, and I am pleased to state that it is filled with examples of the best methods of modern practice. We shall introduce it as a text-book in the college course." This is a fair specimen of the general reception.

Mathematical Almanac and Annual says:—

"Davies is a deservedly popular author, and his mathematical works are text-books in many of the leading schools and colleges."

Van Nostrand's Eclectic Engineering Magazine says:—

"We find in this new work all that can be asked for in a text-book. If there is a better work than this on Surveying, either for students or surveyors, our attention has not been called to it."

THE NEW LEGENDRE.

Van Amringe's Davies' Legendre.

Elements of Geometry and Trigonometry. By Charles Davies, LL.D. Revised (1885) by Prof. J. H. Van Amringe of Columbia College. New pages. 8vo. Full leather.

The present edition of the Legendre is the result of a careful re-examination of the work, into which have been incorporated such emendations in the way of greater clearness of expression or of proof as could be made without altering it in form or substance. Practical exercises are placed at the end of the several books, and comprise additional theorems, problems, and numerical exercises upon the principles of the Book or Books preceding. They will be found of great service in accustoming students, early in and throughout their course, to make for themselves practical application of geometric principles, and constitute, in addition, a large and excellent body of review and test questions for the convenience of teachers. The Trigonometry and mensuration have been carefully revised throughout ; the deduction of principles and rules has been simplified ; the discussion of the several cases which arise in the solution of triangles, plane and spherical, has been made more full and clear ; and the whole has, in definition, demonstration, illustration, &c., been made to conform to the latest and best methods.

It is believed that in clearness and precision of definition, in general simplicity and rigor of demonstration, in the judicious arrangement of practical exercises, in orderly and logical development of the subject, and in compactness of form, Davies' Legendre is superior to any work of its grade for the general training of the logical powers of pupils, and for their instruction in the great body of elementary geometric truth.

The work has been printed from entirely new plates, and no care has been spared to make it a model of typographical excellence

DAVIES'S NATIONAL COURSE OF MATHEMATICS.

ITS RECORD.

In claiming for this series the first place among American text-books, of whatever class, the publishers appeal to the magnificent record which its volumes have earned during the *thirty-five years* of Dr. Charles Davies's mathematical labors. The unremitting exertions of a life-time have placed *the modern series* on the same proud eminence among competitors that each of its predecessors had successively enjoyed in a course of constantly improved editions, now rounded to their perfect fruition, — for it seems almost that this science is susceptible of no further demonstration.

During the period alluded to, many authors and editors in this department have started into public notice, and, by borrowing ideas and processes original with Dr. Davies, have enjoyed a brief popularity, but are now almost unknown. Many of the series of to-day, built upon a similar basis, and described as "modern books," are destined to a similar fate; while the most far-seeing eye will find it difficult to fix the time, on the basis of any data afforded by their past history, when these books will cease to increase and prosper, and fix a still firmer hold on the affection of every educated American.

One cause of this unparalleled popularity is found in the fact that the enterprise of the author did not cease with the original completion of his books. Always a practical teacher, he has incorporated in his text-books from time to time the advantages of every improvement in methods of teaching, and every advance in science. During all the years in which he has been laboring he constantly submitted his own theories and those of others to the practical test of the class-room, approving, rejecting, or modifying them as the experience thus obtained might suggest. In this way he has been able to produce an almost perfect series of class-books, in which every department of mathematics has received minute and exhaustive attention.

Upon the death of Dr. Davies, which took place in 1876, his work was immediately taken up by his former pupil and mathematical associate of many years, Prof. W. G. Peck, LL.D., of Columbia College. By him, with Prof. J. H. Van Amringe, of Columbia College, the original series is kept carefully revised and up to the times.

DAVIES'S SYSTEM IS THE ACKNOWLEDGED NATIONAL STANDARD FOR THE UNITED STATES, for the following reasons: —

1st. It is the basis of instruction in the great national schools at West Point and Annapolis.

2d. It has received the *quasi* indorsement of the National Congress.

3d. It is exclusively used in the public schools of the National Capital.

4th. The officials of the Government use it as authority in all cases involving mathematical questions.

5th. Our great soldiers and sailors commanding the national armies and navies were educated in this system. So have been a majority of eminent scientists in this country All these refer to "Davies" as authority.

6th. A larger number of American citizens have received their education from this than from any other series.

7th. The series has a larger circulation throughout the whole country than any other, being *extensively used in every State in the Union.*

DAVIES AND PECK'S ARITHMETICS.

OPTIONAL OR CONSECUTIVE.

The best thoughts of these two illustrious mathematicians are combined in the following beautiful works, which are the natural successors of Davies's Arithmetics, sumptuously printed, and bound in crimson, green, and gold: —

Davies and Peck's Brief Arithmetic.

Also called the "Elementary Arithmetic." It is the shortest presentation of the subject, and is *adequate* for all grades in common schools, being a thorough introduction to practical life, except for the specialist.

At first the authors play with the little learner for a few lessons, by object-teaching and kindred allurements; but he soon begins to realize that study is earnest, as he becomes familiar with the simpler operations, and is delighted to find himself master of important results.

The second part reviews the Fundamental Operations on a scale proportioned to the enlarged intelligence of the learner. It establishes the General Principles and Properties of Numbers, and then proceeds to Fractions. Currency and the Metric System are fully treated in connection with Decimals. Compound Numbers and Reduction follow, and finally Percentage with all its varied applications.

An Index of words and principles concludes the book, for which every scholar and most teachers will be grateful. How much time has been spent in searching for a half-forgotten definition or principle in a former lesson !

Davies and Peck's Complete Arithmetic.

This work certainly deserves its name in the best sense. Though complete, it is not, like most others which bear the same title, *cumbersome*. These authors excel in clear, lucid demonstrations, teaching the science pure and simple, yet not ignoring convenient methods and practical applications.

For turning out a thorough business man no other work is so well adapted. He will have a clear comprehension of the science as a whole, and a working acquaintance with details which must serve him well in all emergencies. Distinguishing features of the book are the logical progression of the subjects and the great variety of practical problems, not *puzzles*, which are beneath the dignity of educational science. A clear-minded critic has said of Dr. Peck's work that it is free from that juggling with numbers which some authors falsely call "Analysis." A series of Tables for converting ordinary weights and measures into the Metric System appear in the later editions.

PECK'S ARITHMETICS.

Peck's First Lessons in Numbers.

This book begins with pictorial illustrations, and unfolds gradually the science of numbers. It noticeably simplifies the subject by developing the principles of addition and subtraction simultaneously; as it does, also, those of multiplication and division.

Peck's Manual of Arithmetic.

This book is designed especially or those who seek sufficient instruction to carry them successfully through practical life, but have not time for extended study.

Peck's Complete Arithmetic.

This completes the series but is a much briefer book than most of the complete arithmetics, and is recommended not only for what it contains, but also for what is omitted.

It may be said of Dr. Peck's books more truly than of any other series published, that they are clear and simple in definition and rule, and that superfluous matter of every kind has been faithfully eliminated, thus magnifying the working value of the book and saving unnecessary expense of time and labor.

BARNES'S NEW MATHEMATICS.

In this series JOSEPH FICKLIN, Ph. D., Professor of Mathematics and Astronomy in the University of Missouri, has combined all the best and latest results of practical and experimental teaching of arithmetic with the assistance of many distinguished mathematical authors.

Barnes's Elementary Arithmetic.

Barnes's National Arithmetic.

These two works constitute a *complete arithmetical course in two books.*

They meet the demand for text-books that will help students to acquire the greatest amount of useful and practical knowledge of Arithmetic by the smallest expenditure of *time, labor,* and *money.* Nearly every topic in Written Arithmetic is introduced, and its principles illustrated, by exercises in *Oral* Arithmetic. The free use of Equations; the concise method of combining and treating Properties of Numbers; the treatment of Multiplication and Division of Fractions in *two* cases, and then reduced to *one;* Cancellation by the use of the vertical line, especially in Fractions, Interest, and Proportion; the brief, simple, and greatly superior method of working Partial Payments by the "Time Table" and Cancellation; the substitution of formulas to a great extent for rules; the full and practical treatment of the Metric System, &c., indicate their completeness. A *variety* of methods and processes for the *same topic,* which deprive the pupil of the great benefit of doing a part of the *thinking* and *labor* for himself, have been discarded. The statement of principles, definitions, rules, &c., is brief and simple. The illustrations and methods are explicit, direct, and practical. The great number and variety of Examples embody the actual business of the day. The very large amount of matter condensed in so small a compass has been accomplished by economizing every line of space, by rejecting superfluous matter and obsolete terms, and by avoiding the *repetition* of analyses, explanations, and operations in the advanced topics which have been used in the more elementary parts of these books.

AUXILIARIES.

For use in district schools, and for supplying a text-book in advanced work for classes having finished the course as given in the ordinary Practical Arithmetics, the National Arithmetic has been divided and bound separately, as follows: —

Barnes's Practical Arithmetic.

Barnes's Advanced Arithmetic.

In many schools there are classes that for various reasons never reach beyond Percentage. It is just such cases where *Barnes's Practical Arithmetic* will answer a good purpose, at a *price to the pupil* much less than to buy the complete book. On the other hand, classes having finished the ordinary Practical Arithmetic can proceed with the higher course by using *Barnes's Advanced Arithmetic.*

For primary schools requiring simply a table book, and the earliest rudiments forcibly presented through object-teaching and copious illustrations, we have prepared

Barnes's First Lessons in Arithmetic,

which begins with the most elementary notions of numbers, and proceeds, by simple steps, to develop all the fundamental principles of Arithmetic.

Barnes's Elements of Algebra.

This work, as its title indicates, is elementary in its character and suitable for use, (1) in such public schools as give instruction in the Elements of Algebra: (2) in institutions of learning whose courses of study do not include Higher Algebra; (3) in schools whose object is to prepare students for entrance into our colleges and universities. This book will also meet the wants of students of Physics who require some knowledge of

Algebra. The student's progress in Algebra depends very largely upon the proper treatment of the four *Fundamental Operations.* The terms *Addition, Subtraction, Multiplication,* and *Division* in Algebra have a wider meaning than in Arithmetic, and these operations have been so defined as to *include* their arithmetical meaning; so that the beginner is simply called upon to *enlarge* his views of those fundamental operations. Much attention has been given to the explanation of the negative sign, in order to remove the well-known difficulties in the use and interpretation of that sign. Special attention is here called to "A Short Method of Removing Symbols of Aggregation," Art. 76. On account of their importance, the subjects of *Factoring, Greatest Common Divisor,* and *Least Common Multiple* have been treated at greater length than is usual in elementary works. In the treatment of *Fractions,* a method is used which is quite simple, and, at the same time, more general than that usually employed. In connection with *Radical Quantities* the roots are expressed by fractional exponents, for the principles and rules applicable to integral exponents may then be used without modification. The *Equation* is made the chief subject of thought in this work. It is defined near the beginning, and used extensively in every chapter. In addition to this, four chapters are devoted exclusively to the subject of *Equations.* All *Proportions* are equations, and in their treatment as such all the difficulty commonly connected with the subject of Proportion disappears. The chapter on Logarithms will doubtless be acceptable to many teachers who do not require the student to master Higher Algebra before entering upon the study of Trigonometry.

HIGHER MATHEMATICS.

Peck's Manual of Algebra.

Bringing the methods of Bourdon within the range of the Academic Course.

Peck's Manual of Geometry.

By a method purely practical, and unembarrassed by the details which rather confuse than simplify science.

Peck's Practical Calculus.

Peck's Analytical Geometry.

Peck's Elementary Mechanics.

Peck's Mechanics, with Calculus.

The briefest treatises on these subjects now published. Adopted by the great Universities: Yale, Harvard, Columbia, Princeton, Cornell, &c.

Macnie's Algebraical Equations.

Serving as a complement to the more advanced treatises on Algebra, giving special attention to the analysis and solution of equations with numerical coefficients.

Church's Elements of Calculus.

Church's Analytical Geometry.

Church's Descriptive Geometry. With plates. 2 vols.

These volumes constitute the "West Point Course" in their several departments Prof. Church was long the eminent professor of mathematics at West Point Military Academy, and his works are standard in all the leading colleges.

Courtenay's Elements of Calculus.

A standard work of the very highest grade, presenting the most elaborate attainable survey of the subject.

Hackley's Trigonometry.

With applications to Navigation and Surveying, Nautical and Practical Geometry, and Geodesy.

BARNES'S ONE-TERM HISTORY SERIES.

A Brief History of the United States.

This is probably the MOST ORIGINAL SCHOOL-BOOK published for many years, in any department. A few of its claims are the following: —

1. **Brevity**. — The text is complete for grammar school or intermediate classes, in 290 12mo pages, large type. It may readily be completed, if desired, in one term of study.

2. **Comprehensiveness** — Though so brief, this book contains the pith of all the wearying contents of the larger manuals, and a great deal more than the memory usually retains from the latter.

3. **Interest** has been a prime consideration. Small books have heretofore been bare, full of dry statistics, unattractive. This one is charmingly written, replete with anecdote and brilliant with illustration.

4. **Proportion of Events**. — It is remarkable for the discrimination with which the different portions of our history are presented according to their importance. Thus the older works, being already large books when the Civil War took place, give it less space than that accorded to the Revolution.

5. **Arrangement**. — In six epochs, entitled respectively, Discovery and Settlement, the Colonies, the Revolution, Growth of States, the Civil War, and Current Events.

6. **Catch Words**. — Each paragraph is preceded by its leading thought in prominent type, standing in the student's mind for the whole paragraph.

7. **Key Notes**. — Analogous with this is the idea of grouping battles, &c., about some central event, which relieves the sameness so common in such descriptions, and renders each distinct by some striking peculiarity of its own.

8. **Foot-Notes**. — These are crowded with interesting matter that is not strictly a part of history proper. They may be learned or not, at pleasure. They are certain in any event to be read.

9. **Biographies** of all the leading characters are given in full in foot-notes.

10. **Maps**. — Elegant and distinct maps from engravings on copper-plate, and beautifully colored, precede each epoch, and contain all the places named.

11. **Questions** are at the back of the book, to compel a more independent use of the text. Both text and questions are so worded that the pupil must give intelligent answers IN HIS OWN WORDS. "Yes" and "No" will not do.

HISTORY — *Continued.*

12. **Historical Recreations.** — These are additional questions to test the student's knowledge, in review, as: "What trees are celebrated in our history?" "When did a fog save our army?" "What Presidents died in office?" "When was the Mississippi our western boundary?" "Who said, 'I would rather be right than President'?" &c.

13. **The Illustrations,** about seventy in number, are the work of our best artists and engravers, produced at great expense. They are vivid and interesting, and mostly upon subjects never before illustrated in a school-book.

14. **Dates.** — Only the leading dates are given in the text, and these are so associated as to assist the memory, but at the head of each page is the date of the event first mentioned, and at the close of each epoch a summary of events and dates.

15. **The Philosophy of History** is studiously exhibited, the causes and effects of events being distinctly traced and their inter-connection shown.

16. **Impartiality.** — All sectional, partisan, or denominational views are avoided. Facts are stated after a careful comparison of all authorities without the least prejudice or favor.

17. **Index.** — A verbal index at the close of the book perfects it as a work of reference.

It will be observed that the above are all particulars in which School Histories have been signally defective, or altogether wanting. Many other claims to favor it shares in common with its predecessors.

TESTIMONIALS.

From PROF. WM. F. ALLEN, *State University of Wisconsin.*

"Two features that I like *very much* are the *anecdotes* at the foot of the page and the '*Historical Recreations*' in the Appendix. The latter, I think, is quite a *new* feature, and the other is *very* well executed."

From HON. NEWTON BATEMAN, *Superintendent Public Instruction, Illinois.*

"Barnes's One-Term History of the United States is an exceedingly attractive and spirited little book. Its claim to several new and valuable features seems well founded. Under the form of six well-defined epochs, the history of the United States is traced tersely, yet pithily, from the earliest times to the present day. A good map precedes each epoch, whereby the history and geography of the period may be studied together, *as they always should be.* The syllabus of each paragraph is made to stand in such bold relief, by the use of large, heavy type, as to be of much *mnemonic* value to the student. The book is written in a sprightly and piquant style, the interest never flagging from beginning to end, — a rare and difficult achievement in works of this kind."

From HON. ABNER J. PHIPPS, *Superintendent Schools, Lewiston, Maine.*

"Barnes's History of the United States has been used for several years in the Lewiston schools, and has proved a very satisfactory work. I have examined the new edition of it."

From HON. R. K. BUEHRLE, *City Superintendent Schools, Lancaster, Pa.*

"It is the *best* history of the kind I have ever seen."

From T. J. CHARLTON, *Superintendent Public Schools, Vincennes, Ind.*

"We have used it here for six years, and it has given almost perfect satisfaction. . . . The notes in fine print at the bottom of the pages are of especial value."

From PROF. WM. A. MOWRY, *E. & C. School, Providence, R. I.*

"Permit me to express my high appreciation of your book. I wish all textbooks for the young had equal merit."

From HON. A. M. KEILEY, *City Attorney, Late Mayor, and President of the School Board, City of Richmond, Va.*

"I do not hesitate to volunteer to you the opinion that Barnes's History is entitled to the preference in almost every respect that distinguishes a good schoolbook. . . . The narrative generally exhibits the temper of the judge; rarely, if ever, of the advocate."

A Brief History of Ancient Peoples.

With an account of their monuments, literature, and manners. 340 pages 12mo. Profusely illustrated.

In this work the political history, which occupies nearly, if not all, the ordinary school text, is condensed to the salient and essential facts, in order to give room for a clear outline of the literature, religion, architecture, character, habits, &c., of each nation. Surely it is as important to know *something* about Plato as *all* about Cæsar, and to learn how the ancients wrote their books as how they fought their battles.

The chapters on Manners and Customs and the Scenes in Real Life represent the people of history as men and women subject to the same wants, hopes and fears as ourselves, and so bring the distant past near to us. The Scenes, which are intended *only for reading*, are the result of a careful study of the unequalled collections of monuments in the London and Berlin Museums, of the ruins in Rome and Pompeii, and of the latest authorities on the domestic life of ancient peoples. Though intentionally written in a semi-romantic style, they are accurate pictures of what *might* have occurred, and some of them are simple transcriptions of the details sculptured in Assyrian alabaster or painted on Egyptian walls.

HISTORY — *Continued.*

The extracts made from the sacred books of the East are not specimens of their style and teachings, but only gems selected often from a mass of matter, much of which would be absurd, meaningless, and even revolting. It has not seemed best to cumber a book like this with selections conveying no moral lesson.

The numerous cross-references, the abundant dates in parenthesis, the pronunciation of the names in the Index, the choice reading references at the close of each general subject, and the novel Historical Recreations in the Appendix, will be of service to teacher and pupil alike.

Though designed primarily for a text-book, a large class of persons — general readers, who desire to know something about the progress of historic criticism and the recent discoveries made among the resurrected monuments of the East, but have no leisure to read the ponderous volumes of Brugsch, Layard, Grote, Mommsen, and Ihne — will find this volume just what they need.

From HOMER B. SPRAGUE, *Head Master Girls' High School, West Newton St., Boston, Mass.*

"I beg to recommend in strong terms the adoption of Barnes's 'History of Ancient Peoples' as a text-book. It is about as nearly perfect as could be hoped for. The adoption would give great relish to the study of Ancient History."

HE Brief History of France.

By the author of the "Brief United States," with all the attractive features of that popular work (which see) and new ones of its own.

It is believed that the History of France has never before been presented in such brief compass, and this is effected without sacrificing one particle of interest. The book reads like a romance, and, while drawing the student by an irresistible fascination to his task, impresses the great outlines indelibly upon the memory.

HISTORY — *Continued.*

Barnes's Brief History of Mediæval and Modern Peoples.

The success of the History of Ancient Peoples was immediate and great. A History of Mediæval and Modern History, upon the same plan, was the natural sequence. Those teachers who used the former will be glad to know that the latter book is now ready, and classes can go right on without changing authors.

The New York *School Journal* says: — "The fine-print notes . . . work a field not widely developed until Green's History of English People appeared, relating to the description of real, every-day life of the people."

This work distinguishes between the period of the world's history from the Fall of Rome (A.D. 476) to the Capture of Constantinople (A.D. 1453), — about one thousand years, called "Middle Ages," — and the period from the end of the fifteenth century to the present time. It covers the entire time chronologically and by the order of events, giving one hundred and twenty-two fine illustrations and sixteen elaborate maps.

[Illustration from Barnes's Brief-History Series.]

The subject has never before been so interestingly treated in brief compass. The Political History of each nation is first given, then the Manners and Customs of the People. A better idea of the growth of civilization and the changes in the condition of mankind cannot be found elsewhere. The book is fitted for private reading, as well as schools.

HISTORY — *Continued.*

Barnes's Brief General History.

Comprising Ancient, Mediæval, and Modern Peoples.

THE SPECIAL FEATURES OF THIS BOOK ARE AS FOLLOWS: —

The **General History** contains 600 pages. Of this amount, 350 pages are devoted to the political history, and 250 pages to the civilization, manners, and customs, etc. The latter are in separate chapters, and if the time of the teacher is limited, may be omitted. The class can thus take only the political portion when desired. The teacher will have, however, the satisfaction of knowing that, such is the fascinating treatment of the civilization, literature, etc., those chapters will be carefully read by the pupils; and, on the principle that knowledge acquired from love alone is the most vivid, will probably be the best-remembered part of the book. This portion of the book is therefore all clear gain.

The **Black-board Analysis.** See p. 314 as an example of this marked feature.

The exquisite **Illustrations**, unrivalled by any text-book. See pp. 9, 457, and 582, as samples of the 240 cuts contained in this beautiful work.

The peculiar **Summaries**, and valuable lists of **Reading References.** See p. 417.

The numerous and excellent colored **Maps.** These are so full as to answer for an extensive course of collateral reading, and are consequently useful for reference outside of class-work. See pp. 299 and 317.

The Scenes in Real Life, which are the result of a careful study of the collections and monuments in the London, Paris, and Berlin museums, and the latest authorities upon the domestic life of the people of former times. See pp. 38–39. This scene — a Lord of the IVth Dynasty — is mainly a transcription of details to be found painted on the walls of Egyptian tombs.

The chapters on **Civilization** that attempt to give some idea of the Monuments, Arts, Literature, Education, and Manners and Customs of the different nations. See pp. 171, 180, 276, 279, 472, and 514.

The admirable **Genealogical Tables** interspersed throughout the text. See pp. 340 and 494.

The **Foot-Notes** that are packed full of anecdotes, biographies, pleasant information, and suggestive comments. As an illustration of these, take the description of the famous sieges of Haarlem and Leyden, during the Dutch War of Independence, pp. 446 and 448.

The peculiar method of treating **Early Roman History,** by putting in the text the facts as accepted by critics, and, in the notes below, the legends. See pp. 205–6.

The exceedingly useful plan of running collateral history in parallel columns, as for example on p. 361, taken from the Hundred Years' War.

The **Historical Recreations,** so valuable in arousing the interest of a class. See p. xi from the Appendix.

The striking opening of Modern History on pp. 423–4.

The interesting **Style,** that sweeps the reader along as by the fascination of a novel. The pupil insensibly acquires a taste for historical reading, and forgets the tediousness of the ordinary lesson in perusing the thrilling story of the past. See pp. 251–2.

Special attention is called to the chapter entitled **Rise of Modern Nations,** — England, France, and Germany. The characteristic feature in the mediæval history of each of these nations is made prominent. (*a.*) After the Four Conquests of England, the central idea in the growth of that people was the Development of Constitutional Liberty. (*b.*) The feature of French history was the conquest of the great vassals by the king, the triumph of royalty over feudalism, and the final consolidation of the scattered fiefs into one grand monarchy. (*c.*) The characteristic of German history was disunion, emphasized by the lack of a central capital city, and by an elective rather than an hereditary monarchy. The struggle of the Crown with its powerful vassals was the same as in France, but developed no national sentiment, and ended in the establishment of semi-independent dukedoms.

These three thoughts furnish the beginner with as many threads on which to string the otherwise isolated facts of this bewildering period.

GENERAL HISTORY.

Monteith's Youth's History of the United States.

A History of the United States for beginners. It is arranged upon the catechetical plan, with illustrative maps and engravings, review questions, dates in parentheses (that their study may be optional with the younger class of learners), and interesting biographical sketches of all persons who have been prominently identified with the history of our country.

Willard's United States. School and University Editions.

The plan of this standard work is chronologically exhibited in front of the titlepage. The maps and sketches are found useful assistants to the memory; and dates, usually so difficult to remember, are so systematically arranged as in a great degree to obviate the difficulty. Candor, impartiality, and accuracy are the distinguishing features of the narrative portion.

Willard's Universal History. New Edition.

The most valuable features of the "United States" are reproduced in this. The peculiarities of the work are its great conciseness and the prominence given to the chronological order of events. The margin marks each successive era with great distinctness, so that the pupil retains not only the event but its time, and thus fixes the order of history firmly and usefully in his mind. Mrs. Willard's books are constantly revised, and at all times written up to embrace important historical events of recent date. Professor Arthur Gilman has edited the last twenty-five years to 1882.

Lancaster's English History.

By the Master of the Stoughton Grammar School, Boston. The most practical of the "brief books." Though short, it is not a bare and uninteresting outline, but contains enough of explanation and detail to make intelligible the *cause and effect* of events. Their relations to the history and development of the American people is made specially prominent.

Willis's Historical Reader.

Being Collier's Great Events of History adapted to American schools. This rare epitome of general history, remarkable for its charming style and judicious selection of events on which the destinies of nations have turned, has been skilfully manipulated by Professor Willis, with as few changes as would bring the United States into its proper position in the historical perspective. As reader or text-book it has few equals and no superior.

Berard's History of England.

By an authoress well known for the success of her History of the United States. The social life of the English people is felicitously interwoven, as in fact, with the civil and military transactions of the realm.

Ricord's History of Rome.

Possesses the charm of an attractive romance. The fables with which this history abounds are introduced in such a way as not to deceive the inexperienced, while adding materially to the value of the work as a reliable index to the character and institutions, as well as the history of the Roman people.

HISTORY — *Continued.*

Hanna's Bible History.

The only compendium of Bible narrative which affords a connected and chronological view of the important events there recorded, divested of all superfluous detail.

Summary of History; American, French, and English.

A well-proportioned outline of leading events, condensing the substance of the more extensive text-books in common use into a series of statements so brief, that every word may be committed to memory, and yet so comprehensive that it presents an accurate though general view of the whole continuous life of nations.

Marsh's Ecclesiastical History.

Affording the History of the Church in all ages, with accounts of the pagan world during the biblical periods, and the character, rise, and progress of all religions, as well as the various sects of the worshippers of Christ. The work is entirely non-sectarian, though strictly catholic. A separate volume contains carefully prepared questions for class use.

Mill's History of the Ancient Hebrews.

With valuable Chronological Charts, prepared by Professor Edwards of N. Y. This is a succinct account of the chosen people of God to the time of the destruction of Jerusalem. Complete in one volume.

Topical History Chart Book.

By Miss Ida P. Whitcomb. To be used in connection with *any History, Ancient or Modern*, instead of the ordinary blank book for summary. It embodies the names of *contemporary rulers* from the earliest to the present time, with blanks under each, in which the pupil may write the summary of the life of the ruler.

Gilman's First Steps in General History.

A "suggestive outline" of rare compactness. Each country is treated by itself, and the United States receive special attention. Frequent maps, contemporary events in tables, references to standard works for fuller details, and a minute Index constitute the "Illustrative Apparatus." From no other work that we know of can so succinct a view of the world's history be obtained. Considering the necessary limitation of space, the style is surprisingly vivid, and at times even ornate. In all respects a charming, though not the less practical, text-book.

Baker's Brief History of Texas.

Dimitry's History of Louisana.

Alison's Napoleon First.

The history of Europe from 1788 to 1815. By Archibald Alison. Abridged by Edward S. Gould. One vol., 8vo, with appendix, questions, and maps. 550 pages.

Lord's Points of History.

The salient points in the history of the world arranged catechetically for class use or for review and examination of teacher or pupil. By John Lord, LL.D. 12mo, 300 pages.

Carrington's Battle Maps and Charts of the American Revolution.

Topographical Maps and Chronological Charts of every battle, with 3 steel portraits of Washington. 8vo, cloth.

Condit's History of the English Bible.

For theological and historical students this book has an intrinsic value. It gives the history of all the English translations down to the present time, together with a careful review of their influence upon English literature and language.

DRAWING.

BARNES'S POPULAR DRAWING SERIES.

Based upon the experience of the most successful teachers of drawing in the United States.

The Primary Course, consisting of a manual, ten cards, and three primary drawing books, A, B, and C.

Intermediate Course. Four numbers and a manual.

Advanced Course. Four numbers and a manual.

Instrumental Course. Four numbers and a manual.

The Intermediate, Advanced, and Instrumental Courses are furnished either in book or card form at the same prices. The books contain the usual blanks, with the unusual advantage of opening from the pupil, — placing the copy directly in front and above the blank, thus occupying but little desk-room. The cards are in the end more economical than the books, if used in connection with the patent blank folios that accompany this series.

The cards are arranged to be bound (or tied) in the folios and removed at pleasure. The pupil at the end of each number has a complete book, containing only his own work, while the copies are preserved and inserted in another folio ready for use in the next class.

Patent Blank Folios. No. 1. Adapted to Intermediate Course. No. 2. Adapted to Advanced and Instrumental Courses.

ADVANTAGES OF THIS SERIES.

The Plan and Arrangement. — The examples are so arranged that teachers and pupils can see, at a glance, how they are to be treated and where they are to be copied. In this system, copying and designing do not receive all the attention. The plan is broader in its aims, dealing with drawing as a branch of common-school instruction, and giving it a wide educational value.

Correct Methods. — In this system the pupil is led to rely upon himself, and not upon delusive mechanical aids, as printed guide-marks, &c.

One of the principal objects of any good course in freehand drawing is to educate the eye to estimate location, form, and size. A system which weakens the motive or removes the necessity of *thinking* is false in theory and ruinous in practice. The object should be to educate, not cram; to develop the intelligence, not teach tricks.

Artistic Effect. — The beauty of the examples is not destroyed by crowding the pages with useless and badly printed text. The Manuals contain all necessary instruction.

Stages of Development. — Many of the examples are accompanied by diagrams, showing the different stages of development.

Lithographed Examples. — The examples are printed in imitation of pencil drawing (not in hard, black lines) that the pupil's work may resemble them.

One Term's Work. — Each book contains what can be accomplished in an average term, and no more. Thus a pupil *finishes* one book before beginning another.

Quality — not Quantity. — Success in drawing depends upon the amount of *thought* exercised by the pupil, and *not* upon the large number of examples drawn.

Designing. — Elementary design is more skilfully taught in this system than by any other. In addition to the instruction given in the books, the pupil will find printed on the insides of the covers a variety of beautiful patterns.

Enlargement and Reduction. — The practice of enlarging and reducing from copies is not commenced until the pupil is well advanced in the course and therefore better able to cope with this difficult feature in drawing.

Natural Forms. — This is the only course that gives at convenient intervals easy and progressive exercises in the drawing of natural forms.

Economy. — By the patent binding described above, the copies need not be thrown aside when a book is filled out, but are preserved in perfect condition for future use. The blank books, only, will have to be purchased after the first introduction, thus effecting a saving of more than half in the usual cost of drawing-books.

Manuals for Teachers. — The Manuals accompanying this series contain practical instructions for conducting drawing in the class-room, with *definite* directions for drawing *each* of the examples in the books, instructions for designing, model and object drawing, drawing from natural forms, &c.

DRAWING — *Continued.*

Chapman's American Drawing-Book.

The standard American text-book and authority in all branches of art A compilation of art principles. A manual for the amateur, and basis of study for the professional artist. Adapted for schools and private instruction.

CONTENTS. — "Any one who can Learn to Write can Learn to Draw." — Primary Instruction in Drawing. — Rudiments of Drawing the Human Head. — Rudiments in Drawing the Human Figure. — Rudiments of Drawing. — The Elements of Geometry. — Perspective. — Of Studying and Sketching from Nature. — Of Painting. — Etching and Engraving. — Of Modelling. — Of Composition. — Advice to the American Art-Student. The work is of course magnificently illustrated with all the original designs.

Chapman's Elementary Drawing-Book.

A progressive course of practical exercises, or a text-book for the training of the eye and hand. It contains the elements from the larger work, and a copy should be in the hands of every pupil; while a copy of the "American Drawing-Book," named above, should be at hand for reference by the class.

Clark's Elements of Drawing.

A complete course in this graceful art, from the first rudiments of outline to the finished sketches of landscape and scenery.

Allen's Map-Drawing and Scale.

This method introduces a new era in map-drawing, for the following reasons: 1. It is a system. This is its greatest merit. — 2. It is easily understood and taught. — 3. The eye is trained to exact measurement by the use of a scale. — 4. By no special effort of the memory, distance and comparative size are fixed in the mind. — 5. It discards useless construction of lines. — 6. It can be taught by any teacher, even though there may have been no previous practice in map-drawing. — 7. Any pupil old enough to study geography can learn by this system, in a short time, to draw accurate maps. — 8. The system is not the result of theory, but comes directly from the school-room. It has been thoroughly and successfully tested there, with all grades of pupils. — 9. It is economical, as it requires no mapping plates. It gives the pupil the ability of rapidly drawing accurate maps.

FINE ARTS.

Hamerton's Art Essays (Atlas Series): —

No. 1. The Practical Work of Painting.
With portrait of Rubens. 8vo. Paper covers.

No. 2. Modern Schools of Art.
Including American, English, and Continental Painting. 8vo. Paper covers.

Huntington's Manual of the Fine Arts.

A careful manual of instruction in the history of art, up to the present time.

Boyd's Kames' Elements of Criticism.

The best edition of the best work on art and literary criticism ever produced in English.

Benedict's Tour Through Europe.

A valuable companion for any one wishing to visit the galleries and sights of the continent of Europe, as well as a charming book of travels.

Dwight's Mythology.

A knowledge of mythology is necessary to an appreciation of ancient art.

Walker's World's Fair.

The industrial and artistie display at the Centennial Exhibition.

DR. STEELE'S ONE-TERM SERIES, IN ALL THE SCIENCES.

Steele's 14-Weeks Course in Chemistry.
Steele's 14-Weeks Course in Astronomy.
Steele's 14-Weeks Course in Physics.
Steele's 14-Weeks Course in Geology.
Steele's 14-Weeks Course in Physiology.
Steele's 14-Weeks Course in Zoölogy.
Steele's 14-Weeks Course in Botany.

Our text-books in these studies are, as a general thing, dull and uninteresting. They contain from 400 to 600 pages of dry facts and unconnected details. They abound in that which the student cannot learn, much less remember. The pupil commences the study, is confused by the fine print and coarse print, and neither knowing exactly what to learn nor what to hasten over, is crowded through the single term generally assigned to each branch, and frequently comes to the close without a definite and exact idea of a single scientific principle.

Steele's "Fourteen-Weeks Courses" contain only that which every well-informed person should know, while all that which concerns only the professional scientist is omitted. The language is clear, simple, and interesting, and the illustrations bring the subject within the range of home life and daily experience. They give such of the general principles and the prominent facts as a pupil can make familiar as household words within a single term. The type is large and open; there is no fine print to annoy; the cuts are copies of genuine experiments or natural phenomena, and are of fine execution.

In fine, by a system of condensation peculiarly his own, the author reduces each branch to the limits of a single term of study, while sacrificing nothing that is essential, and nothing that is usually retained from the study of the larger manuals in common use. Thus the student has rare opportunity to *economize his time*, or rather to employ that which he has to the best advantage.

A notable feature is the author's charming "style," fortified by an enthusiasm over his subject in which the student will not fail to partake. Believing that Natural Science is full of fascination, he has moulded it into a form that attracts the attention and kindles the enthusiasm of the pupil.

The recent editions contain the author's "Practical Questions" on a plan never before attempted in scientific text-books. These are questions as to the nature and cause of common phenomena, and are not directly answered in the text, the design being to test and promote an intelligent use of the student's knowledge of the foregoing principles.

Steele's Key to all His Works.

This work is mainly composed of answers to the Practical Questions, and solutions of the problems, in the author's celebrated "Fourteen-Weeks Courses" in the several sciences, with many hints to teachers, minor tables, &c. Should be on every teacher's desk.

Prof. J. Dorman Steele is an indefatigable student, as well as author, and his books have reached a fabulous circulation. It is safe to say of his books that they have accomplished more tangible and better results in the class-room than any other ever offered to American schools, and have been translated into more languages for foreign schools. They are even produced in raised type for the blind.

THE NEW GANOT.

Introductory Course of Natural Philosophy.

This book was originally edited from Ganot's "Popular Physics," by William G. Peck, LL.D., Professor of Mathematics and Astronomy, Columbia College, and of Mechanics in the School of Mines. It has recently been revised by Levi S. Burbank, A. M., late Principal of Warren Academy, Woburn, Mass., and James I. Hanson, A.M., Principal of the High School, Woburn, Mass.

Of elementary works those of M. Ganot stand pre-eminent, not only as popular treatises, but as thoroughly scientific expositions of the principles of Physics. His "Traité de Physique" has not only met with unprecedented success in France, but has been extensively used in the preparation of the best works on Physics that have been issued from the American press.

In addition to the "Traité de Physique," which is intended for the use of colleges and higher institutions of learning, M. Ganot published this more elementary work, adapted to the use of schools and academies, in which he faithfully preserved the prominent features and all the scientific accuracy of the larger work. It is charcterized by a well-balanced distribution of subjects, a logical development of scientific principles, and a remarkable clearness of definition and explanation. In addition, it is profusely illustrated with beautifully executed engravings, admirably calculated to convey to the mind of the student a clear conception of the principles unfolded. Their completeness and accuracy are such as to enable the teacher to dispense with much of the apparatus usually employed in teaching the elements of Physical Science.

After several years of great popularity the American publishers have brought this important book thoroughly up to the times. The death of the accomplished educator, Professor Burbank, took place before he had completed his work, and it was then taken in hand by his friend, Professor Hanson, who was familiar with his plans, and has ably and satisfactorily brought the work to completion.

The essential characteristics and general plan of the book have, so far as possible, been retained, but at the same time many parts have been entirely rewritten, much new matter added, a large number of new cuts introduced, and the whole treatise thoroughly revised and brought into harmony with the present advanced stage of scientific discovery.

Among the new features designed to aid in teaching the subject-matter are the summaries of topics, which, it is thought, will be found very convenient in short reviews.

As many teachers prefer to prepare their own questions on the text, and many do not have time to spend in the solution of problems, it has been deemed expedient to insert both the review questions and problems at the end of the volume, to be used or not at the discretion of the instructor.

From the Churchman.

"No department of science has undergone so many improvements and changes in the last quarter of a century as that of natural philosophy. So many and so important have been the discoveries and inventions in every branch of it that everything seems changed but its fundamental principles. Ganot has chapter upon chapter upon subjects that were not so much as known by name to Olmsted; and here we have Ganot, first edited by Professor Peck, and afterward revised by the late Mr. Burbank and Mr. Hanson. No elementary works upon philosophy have been superior to those of Ganot, either as popular treatises or as scientific expositions of the principles of physics, and his 'Traité de Physique' has not only had a great success in France, but has been freely used in this country in the preparation of American books upon the subjects of which it treats. That work was intended for higher institutions of learning, and Mr. Ganot prepared a more elementary work for schools and academies. It is as scientifically accurate as the larger work, and is characterized by a logical development of scientific principles, by clearness of definition and explanation, by a proper distribution of subjects, and by its admirable engravings. We here have Ganot's work enhanced in value by the labors of Professor Peck and of Messrs. Burbank and Hanson, and brought up to our own times. The essential characteristics of Ganot's work have been retained, but much of the book has been rewritten, and many new cuts have been introduced, made necessary by the progress of scientific discovery. The short reviews, the questions on the text, and the problems given for solution are desirable additions to a work of this kind, and will give the book increased popularity.'

FAMILIAR SCIENCE.

Norton & Porter's First Book of Science.

Sets forth the principles of Natural Philosophy, Astronomy, Chemistry, Physiology, and Geology, on the catechetical plan for primary classes and beginners.

Chambers's Treasury of Knowledge.

Progressive lessons upon — *first*, common things which lie most immediately around us, and first attract the attention of the young mind; *second*, common objects from the mineral, animal, and vegetable kingdoms, manufactured articles, and miscellaneous substances; *third*, a systematic view of nature under the various sciences. May be used as a reader or text-book.

Monteith's Easy Lessons in Popular Science.

This book combines within its covers more attractive features for the study of science by children than any other book published. It is a reading book, spelling book, composition book, drawing book, geography, history, book on botany, zoölogy, agriculture, manufactures, commerce, and natural philosophy. All these subjects are presented in a simple and effective style, such as would be adopted by a good teacher on an excursion with a class. The class are supposed to be taking excursions, with the help of a large pictorial chart of geography, which can be suspended before them in the school-room. A key of the chart is inserted in every copy of the book. With this book the science of common or familiar things can be taught to beginners.

NATURAL PHILOSOPHY.

Norton's First Book in Natural Philosophy.

Peck's Elements of Mechanics.

A suitable introduction to Bartlett's higher treatises on Mechanical Philosophy, and adequate in itself for a complete academical course.

Bartlett's Analytical Mechanics.

Bartlett's Acoustics and Optics.

A complete system of Collegiate Philosophy, by Prof. W. H. C. Bartlett, of West Point Military Academy.

Steele's Physics.

Peck's Ganot.

GEOLOGY.

Page's Elements of Geology.

A volume of Chambers's Educational Course. Practical, simple, and eminently calculated to make the study interesting.

Steele's Geology.

CHEMISTRY.

Porter's First Book of Chemistry.

Porter's Principles of Chemistry.

The above are widely known as the productions of one of the most eminent scientific men of America. The extreme simplicity in the method of presenting the science, while exhaustively treated, has excited universal commendation.

Gregory's Chemistry (Organic and Inorganic). 2 vols.

The science exhaustively treated. For colleges and medical students.

Steele's Chemistry.

NATURAL SCIENCE — *Continued.*

ASTRONOMY.

Peck's Popular Astronomy.

By Wm. G. Peck, Ph.D., LL.D., Professor of Mathematics, Mechanics, and Astronomy in Columbia College. 12mo. Cloth. 330 pages.

Professor Peck has here produced a scientific work in brief form for colleges, academies, and high schools. Teachers who do not want an elementary work — like Steele's Astronomy, for instance — will find what they want in this book. Its discussion of the Stars, Solar System, Earth, Moon, Sun and Planets, Eclipses, Tides, Calendars, Planets and Satellites, Comets and Meteors, &c., is full and satisfactory. The illustrations are numerous and very carefully engraved, so the student can gain an accurate comprehension of the things represented. Professor Peck is wonderfully clear and concise in his style of writing, and there is nothing redundant or obscure in this work. It is intended for popular as well as class use, and accordingly avoids too great attention to mathematical processes, which are introduced in smaller type than the regular text. For higher schools this astronomy is undoubtedly the best text-book yet published.

Willard's School Astronomy.

By means of clear and attractive illustrations, addressing the eye in many cases by analogies, careful definitions of all necessary technical terms, a careful avoidance of verbiage and unimportant matter, particular attention to analysis, and a general adoption of the simplest methods, Mrs. Willard has made the best and most attractive *elementary* Astronomy extant.

McIntyre's Astronomy and the Globes.

A complete treatise for intermediate classes. Highly approved.

Bartlett's Spherical Astronomy.

The West Point Course, for advanced classes, with applications to the current wants of Navigation, Geography, and Chronology.

NATURAL HISTORY.

Carll's Child's Book of Natural History.

Illustrating the animal, vegetable, and mineral kingdoms, with application to the arts. For beginners. Beautifully and copiously illustrated.

Anatomical Technology. Wilder & Gage.

As applied to the domestic cat. For the use of students of medicine.

ZOÖLOGY.

Chambers's Elements of Zoölogy.

A complete and comprehensive system of Zoölogy, adapted for academic instruction, presenting a systematic view of the animal kingdom as a portion of external nature.

ROADS AND RAILROADS.

Gillespie's Roads and Railroads.

Tenth Edition. Edited by Cady Staley, A.M., C.E. 464 pages. 12mo. Cloth.

This book has long been and still is the standard manual of the principles and practice of Road-making, comprising the location, construction, and improvement of roads (common, macadam, paved, plank, &c.) and railroads. It was compiled by Wm. Gillespie, LL.D., C.E., of Union College.

LITERATURE.

Gilman's First Steps in English Literature.

The character and plan of this exquisite little text-book may be best understood from an analysis of its contents: Introduction. Historical Period of Immature English, with Chart; Definition of Terms; Languages of Europe, with Chart; Period of Mature English, with Chart; a Chart of Bible Translations, a Bibliography or Guide to General Reading, and other aids to the student.

Cleveland's Compendiums. 3 vols. 12mo.

ENGLISH LITERATURE. AMERICAN LITERATURE.
ENGLISH LITERATURE OF THE XIXTH CENTURY.

In these volumes are gathered the cream of the literature of the English-speaking people for the school-room and the general reader. Their reputation is national. More than 125,000 copies have been sold.

Boyd's English Classics. 6 vols. Cloth. 12mo.

MILTON'S PARADISE LOST. THOMSON'S SEASONS.
YOUNG'S NIGHT THOUGHTS. POLLOK'S COURSE OF TIME.
COWPER'S TASK, TABLE TALK, &C. LORD BACON'S ESSAYS.

This series of annotated editions of great English writers in prose and poetry is designed for critical reading and parsing in schools. Prof. J. R. Boyd proves himself an editor of high capacity, and the works themselves need no encomium. As auxiliary to the study of belles-lettres, &c., these works have no equal.

Pope's Essay on Man. 16mo. Paper.

Pope's Homer's Iliad. 32mo. Roan.

The metrical translation of the great poet of antiquity, and the matchless "Essay on the Nature and State of Man," by Alexander Pope, afford superior exercise in literature and parsing.

POLITICAL ECONOMY.

Champlin's Lessons on Political Economy.

An improvement on previous treatises, being shorter, yet containing everything essential, with a view of recent questions in finance, &c., which is not elsewhere found.

ÆSTHETICS.

Huntington's Manual of the Fine Arts.

A view of the rise and progress of art in different countries, a brief account of the most eminent masters of art, and an analysis of the principles of art. It is complete in itself, or may precede to advantage the critical work of Lord Kames.

Boyd's Kames's Elements of Criticism.

The best edition of this standard work; without the study of which none may be considered proficient in the science of the perceptions. No other study can be pursued with so marked an effect upon the taste and refinement of the pupil.

ELOCUTION.

Watson's Practical Elocution.

A scientific presentment of accepted principles of elocutionary drill, with blackboard diagrams and full collection of examples for class drill. Cloth. 90 pages, 12mo.

Taverner Graham's Reasonable Elocution.

Based upon the belief that true elocution is the right interpretation of thought, and guiding the student to an intelligent appreciation, instead of a merely mechanical knowledge, of its rules.

Zachos's Analytic Elocution.

All departments of elocution — such as the analysis of the voice and the sentence, phonology, rhythm, expression, gesture, &c. — are here arranged for instruction in classes, illustrated by copious examples.

SPEAKERS.

Northend's Little Orator.

Northend's Child's Speaker.

Two little works of the same grade but different selections, containing simple and attractive pieces for children under twelve years of age.

Northend's Young Declaimer.

Northend's National Orator.

Two volumes of prose, poetry, and dialogue, adapted to intermediate and grammar classes respectively.

Northend's Entertaining Dialogues.

Extracts eminently adapted to cultivate the dramatic faculties, as well as entertain.

Oakey's Dialogues and Conversations.

For school exercises and exhibitions, combining useful instruction.

James's Southern Selections, for Reading and Oratory.

Embracing exclusively Southern literature.

Swett's Common School Speaker.

Raymond's Patriotic Speaker.

A superb compilation of modern eloquence and poetry, with original dramatic exercises. Nearly every eminent modern orator is represented.

MIND.

Mahan's Intellectual Philosophy.

The subject exhaustively considered. The author has evinced learning, candor, and independent thinking.

Mahan's Science of Logic.

A profound analysis of the laws of thought. The system possesses the merit of being intelligible and self-consistent. In addition to the author's carefully elaborated views, it embraces results attained by the ablest minds of Great Britain, Germany, and France, in this department.

Boyd's Elements of Logic.

A systematic and philosophic condensation of the subject, fortified with additions from Watts, Abercrombie, Whately, &c.

Watts on the Mind. Edited by Stephen N. Fellows.

The "Improvement of the Mind," by Isaac Watts, is designed as a guide for the attainment of useful knowledge. As a text-book it is unparalleled; and the discipline it affords cannot be too highly esteemed by the educator.

MORALS.

Peabody's Moral Philosophy.

A short course, by the Professor of Christian Morals, Harvard University, for the Freshman class and for high schools.

Butler's Analogy. Hobart's Analysis.

Edited by Prof. Charles E. West, of Brooklyn Heights Seminary. 228 pages. 16mo. Cloth.

Alden's Text-Book of Ethics.

For young pupils. To aid in systematizing the ethical teachings of the Bible, and point out the coincidences between the instructions of the sacred volume and the sound conclusions of reason.

Smith's Elements of Moral Philosophy.

140 pages. 12mo. Cloth. By Wm. Austin Smith, A.M., Ph.D., Professor of Moral Philosophy in the Columbia (Tenn.) Athenæum.

This is an excellent book for the use of academies and schools. It is prepared to meet the wants of a much larger public than has heretofore been reached by works of this class. The subject is presented in clear and simple language, and will be found adapted to the comprehension of young pupils, at a time when they particularly need an insight into the laws which govern the moral world.

Janet's Elements of Morals.

By Paul Janet. Translated by Mrs. Prof. Corson, of Cornell University.

The **Elements of Morals** is one of a series of works chiefly devoted to Ethics, and treats of practical, rather than theoretical morality.

Mr. Janet is too well known that it be necessary to call attention to his excellence as a moral writer, and it will be sufficient to say that what particularly recommends the **Elements of Morals** to educators and students in general is the admirable adaptation of the book to college and school purposes.

Besides the systematic and scholarly arrangement of its parts, it contains series of examples and illustrations — anecdotic, historical — gathered with rare impartiality from both ancient and modern writers, and which impart a peculiar life and interest to the subject.

Another feature of the work is its sound religious basis. Mr. Janet is above all a religious moralist.

GOVERNMENT.

Young's Lessons in Civil Government.

A comprehensive view of Government, and abstract of the laws showing the rights, duties, and responsibilities of citizens.

Mansfield's Political Manual.

This is a complete view of the theory and practice of the General and State Governments, designed as a text-book. The author is an esteemed and able professor of constitutional law, widely known for his sagacious utterances in the public press.

Martin's Civil Government.

From Prof. Geo. B. Emerson, Boston.

"It is clear and well arranged, and very comprehensive. Whoever reads it attentively will understand more fully and satisfactorily than he could have done without it the history of his own country, and any other. Every young man should study it before he comes to vote, and it should therefore be a text-book in every High School and Academy, and a part of the library of every lover of his country."

From F. P. Conn, Co. Supt. of Schools, Vanderburgh Co., Ind.

"It embraces the essential knowledge of the science, and its arrangement affords ready references to a contents easily acquired. Am satisfied that no more useful book could be adopted, especially in the ungraded schools of the country, where libraries and newspapers are rare."

Antebellum Constitutions.

A complete collection of State and Federal Constitutions as they stood before the Civil War of 1861. With an essay on changes made during the reconstruction period, by Wilmot L. Warren.

PUNCTUATION.

Cocker's Handbook of Punctuation.

With instructions for capitalization, letter-writing, and proof-reading. Most works on this subject are so abstruse and technical that the unprofessional reader finds them difficult of comprehension; but this little treatise is so simple and comprehensive that persons of very ordinary intelligence can readily understand and apply its principles.

ANATOMY.

Anatomical Technology as Applied to the Domestic Cat.

An introduction to human, veterinary, and comparative anatomy. A practical work for students and teachers. 600 pages. 130 figures, and four lithograph plates. By Burt G. Wilder and Simon H. Gage, Professors in Cornell University.

"Instructions in the best method of dissection and study of each organ and region." —*American Veterinary Review.*

"A valuable manual, at once authoritative in statement and admirable in method." —*American Journal of Medical Science.*

"Well adapted to the purpose for which it has been written." —*Nature.*

"The student who will carefully dissect a few cats according to the rules given in this book will have a great advantage over the one who begins his work with the human body; and if he will master the instructions for the various methods of preparation, he will know more than most graduates in medicine." —*The Boston Medical and Surgical Journal.*

FRENCH.

Worman's First French Book.

On same plan as the German and Spanish. The scholar reads and speaks from the first hour understandingly and accurately. 83 pages.

Worman's Second French Book.

Continues the work of the First Book, and is a valuable Elementary French Reader. 96 pages.

Worman's Le Questionnaire.

Exercises on the First French Book. 98 pages. Cloth.

Worman's Grammaire Française.

Written in simple French, but based on English analogy. It therefore dwells upon *the Essentials*, especially those which point out *the variations* of the French from the student's vernacular. 184 pp.

Worman's Teacher's Hand-Book.

Or Key to the Grammaire Française.

Worman's French Echo.

This is not a mass of meaningless and parrot-like phrases thrown together for a tourist's use, to bewilder him when in the presence of a Frenchman.

The "Echo de Paris" is a *strictly progressive conversational book*, beginning with simple phrases and leading by frequent repetition to a mastery of the *idioms* and of *the every-day language* used in business, on travel, at a hotel, in the chit-chat of society.

It presupposes an elementary knowledge of the language, such as may be acquired from the First French Book by Professor Worman, and furnishes *a running French text*, allowing the learner of course to find the meaning of the words (in the appended Vocabulary), and forcing him, by the absence of English in the text, to *think in French.*

Cher Monsieur Worman, — Vous me demandez mon opinion sur votre "Echo de Paris" et quel usage j'en fais. Je ne saurais mieux vous répondre qu'en reproduisant une lettre que j'écrivais dernièrement à un collègue qui était, me disait-il, "bien fatigué de ces insipides livres de dialogues."

"Vous ne connaissez donc pas," lui disais-je, "'l'Echo de Paris,' édité par le Professor Worman? C'est un véritable trésor, merveilleusement adapté au développement de la conversation familière et pratique, telle qu'on la veut aujourd'hui. Cet excellent livre met successivement en scène, d'une manière vive et intéressante, *toutes* les circonstances possibles de la vie ordinaire. Voyez l'immense avantage il vous transporte en France; du premier mot, je m'imagine, et mes élèves avec moi, que nous sommes à Paris, dans la rue, sur une place, dans une gare, dans un salon, dans une chambre, voire même à la cuisine; je parle comme avec des Français; les élèves ne songent pas à traduire de l'anglais pour me répondre; ils pensent en français; ils sont Français pour le moment par les yeux, par l'oreille, par la pensée Quel autre livre pourrait produire cette illusion? . . ."

Votre tout dévoué,

A. de Rougemont.

Illustrated Language Primers.

French and English. German and English.

Spanish and English.

The names of common objects properly illustrated and arranged in easy lessons.

Pujol's Complete French Class-Book.

Offers in one volume, methodically arranged, a complete French course — usually embraced in series of from five to twelve books, including the bulky and expensive lexicon. Here are grammar, conversation, and choice literature, selected from the best French authors. Each branch is thoroughly handled; and the student, having diligently completed the course as prescribed, may consider himself, without further application, *au fait* in the most polite and elegant language of modern times.

MODERN LANGUAGES—*Continued.*

Pujol's French Grammar, Exercises, Reader. 3 vols.

These volumes contain Part I., Parts II. and III., and Part IV. o. the Complete Class-Book respectively, for the convenience of scholars and teachers. The Lexicon is bound with each part.

Maurice-Poitevin's Grammaire Française.

American schools are at last supplied with an American edition of this famous text-book. Many of our best institutions have for years been procuring it from abroad rather than forego the advantages it offers. The policy of putting students who have acquired some proficiency from the ordinary text-books, into a Grammar written in the vernacular, cannot be too highly commended. It affords an opportunity for finish and review at once, while embodying abundant practice of its own rules.

SPANISH.

Worman's First Spanish Book.

On same plan as Worman's first German and French Books. Teaches by direct appeal to illustrations, and by contrast, association, and natural inference. 96 pp.

These little books work marvels in the school-room. The exercises are so developed out of pictured objects and actions, and are so well graduated, that almost from the very outset they go alone. A beginner would have little use for a dictionary in reading. The words are so introduced, and so often used, that the meaning is kept constantly before the mind, without the intervention of a translation.

OTHER SPANISH BOOKS TO FOLLOW.

ANCIENT LANGUAGES.

LATIN.

Searing's Virgil's Æneid, Georgics, and Bucolics.

1. It contains the first six books of the Æneid and the entire Bucolics and Georgics. 2. A very carefully constructed Dictionary. 3. Sufficiently copious notes. 4. Grammatical references to four leading Grammars. 5. Numerous illustrations of the highest order. 6. A superb map of the Mediterranean and adjacent countries. 7. Dr. S. H. Taylor's "Questions on the Æneid." 8. A Metrical Index, and an essay on the Poetical Style. 9. A photographic *fac-simile* of an early Latin MS. 10. The text is according to Jahn, but paragraphed according to Ladewig. 11. Superior mechanical execution.

"My attention was called to Searing's Virgil by the fact of its containing a vocabulary which would obviate the necessity of procuring a lexicon. But use in the class-room has impressed me most favorably with the accuracy and just proportion of its notes, and the general excellence of its grammatical suggestions. The general character of the book, in its paper, its typography, and its engravings, is highly commendable, and the *fac-simile* manuscript is a valuable feature. I take great pleasure in commending the book to all who do not wish a complete edition of Virgil. It suits our short school courses admirably." HENRY L. BOLTWOOD, *Master Princeton High School, Ill.*

Johnson's Persius.

The Satires of Aulus Persius Flaccus, edited, with English notes, principally from Conington. By Henry Clark Johnson, A. M., LL.B., Professor of Latin in the Lehigh University.

CHARTS, &c. — *Continued.*

Popular Folding Reading Charts.

In two parts. Price $5.00 each. These fifty-three charts are the outgrowth of practical reading lessons, all of which have been tried with classes of little children, **first** as black-board lessons, and afterward as printed manuscripts. By this method all the lessons were adapted to the capacity of the children. The words have been carefully selected and graded from the child's own spoken vocabulary.

PART I.

The new words of the first part are taught by the word and sentence method, the object-words being illustrated by engravings.

All the lessons sparkle with real childlike expressions. The language is the language of childhood, and thus to the pupil becomes doubly interesting while at the same time progressive.

The **Clock Face,** with Movable Hands, is an important and attractive feature. The authors know from experience that very happy results can be had by its use. Teaching children to tell the time has always been expected of the teacher, though seldom, if ever, has an opportunity been afforded him to do so.

All the letters of the alphabet are taught by a **series of writing lessons** in the order of their development, and are finally grouped together in a script alphabet.

PART II.

takes up the development of the **elementary sounds** of the language, from the words already learned in Part I., in such a way as to enable the child to see for himself how words are made, and giving the key by which he can make out for himself new words.

A series of **language lessons** is the feature of this part, by which children are gradually taught the use of words by composing brief sentences and original stories.

The **Color Chart** is the most unique feature ever offered to the public, enabling the teacher to teach the primary and secondary colors from nature.

Many review lessons are given in order that the children may learn to read **by reading.**

No easel or framework of any kind is required with the chart. The publishers have secured the exclusive right to use **Shepard's Patent Chart Binding,** the use of which gives it a decided advantage over any other reading chart yet made. It is in this respect unapproachable.

APPARATUS.

Bock's Physiological Apparatus.

A collection of twenty-seven anatomical models.

Harrington's Fractional Blocks.

Harrington's Geometrical Blocks.

These patent blocks are *hinged*, so that each form can be dissected.

Kendall's Lunar Telluric Globe.

Moon, globe, and tellurian combined.

Steele's Chemical Apparatus.

Steele's Geological Cabinet.

Steele's Philosophical Apparatus.

Wood's Botanical Apparatus.

RECORDS.

Cole's Self-Reporting Class Book.

For saving the teacher's labor in averaging. At each opening are a full set of tables showing any scholar's standing at a glance, and entirely obviating the necessity of computation.

Tracy's School Record. {Desk edition. / Pocket edition.}

For keeping a simple but exact record of attendance, deportment, and scholarship. The larger edition contains also a calendar, an extensive list of topics for compositions and colloquies, themes for short lectures, suggestions to young teachers, &c.

Benet's Individual Records.

Brooks's Teacher's Register.

Presents at one view a record of attendance, recitations, and deportment for the whole term.

Carter's Record and Roll-Book.

This is the most complete and convenient record offered to the public. Besides the usual spaces for general scholarship, deportment, attendance, &c., for each name and day, there is a space in red lines enclosing six minor spaces in blue for recording recitations.

National School Diary.

A little book of blank forms for weekly report of the standing of each scholar, from teacher to parent. A great convenience.

REWARDS.

National School Currency.

A little box containing certificates in the form of money. The most entertaining and stimulating system of school rewards. The scholar is paid for his merits and fined for his short-comings. Of course the most faithful are the most successful in business. In this way the use and value of money and the method of keeping accounts are also taught. One box of currency will supply a school of fifty pupils.

LIBRARY AND MISCELLANEOUS PUBLICATIONS.

TEACHERS' WORKING LIBRARY.

Object Lessons. Welch.

This is a complete exposition of the popular modern system of "object-teaching," for teachers of primary classes.

Theory and Practice of Teaching. Page.

This volume has, without doubt, been read by two hundred thousand teachers, and its popularity remains undiminished, large editions being exhausted yearly. It was the pioneer, as it is now the patriarch, of professional works for teachers.

The Graded School. Wells.

The proper way to organize graded schools is here illustrated. The author has availed himself of the best elements of the several systems prevalent in Boston, New York, Philadelphia, Cincinnati, St. Louis, and other cities.

The Normal. Holbrook.

Carries a working school on its visit to teachers, showing the most approved methods of teaching all the common branches, including the technicalities, explanations, demonstrations, and definitions introductory and peculiar to each branch.

School Management. Holbrook.

Treating of the teacher's qualifications; how to overcome difficulties in self and others; organization; discipline; methods of inciting diligence and order; strategy in management; object-teaching.

The Teachers' Institute. Fowle.

This is a volume of suggestions inspired by the author's experience at institutes, in the instruction of young teachers. A thousand points of interest to this class are most satisfactorily dealt with.

Schools and Schoolmasters. Dickens.

Appropriate selections from the writings of the great novelist.

The Metric System. Davies.

Considered with reference to its general introduction, and embracing the views of John Quincy Adams and Sir John Herschel.

The Student; The Educator. Phelps. 2 vols.

The Discipline of Life. Phelps.

The authoress of these works is one of the most distinguished writers on education, and they cannot fail to prove a valuable addition to the School and Teachers' Libraries, being in a high degree both interesting and instructive.

Law of Public Schools. Burke.

By Finley Burke, Counsellor-at-Law. A new volume in "Barnes's Teachers' Library Series." 12mo, cloth.

"Mr. Burke has given us the latest expositions of the law on this highly important subject. I shall cordially commend his treatise." —THEODORE DWIGHT, LL.D.

From the HON. JOSEPH M. BECK, *Judge of Supreme Court, Iowa.*

"I have examined with considerable care the manuscript of 'A Treatise on the Law of Public Schools.' by Finley Burke, Esq., of Council Bluffs. In my opinion, the work will be of great value to school teachers and school officers, and to lawyers. The subjects treated of are thoughtfully considered and thoroughly examined, and correctly and systematically arranged. The style is perspicuous. The legal doctrines of the work, so far as I have been able to consider them, are sound. I have examined quite a number of the authorities cited; they sustain the rules announced in the text. Mr. Burke is an able and industrious member of the bar of the Supreme Court of this State, and has a high standing in the profession of the law."

"I fully concur in the opinion of Judge Beck, above expressed." — JOHN F. DILLON. *New York, May*, 1880.

SIOUX CITY, IOWA, May, 1880.

I have examined the manuscript of Finley Burke, Esq., and find a full citation of all the cases and decisions pertaining to the school law, occurring in the courts of the United States. This volume contains valuable and important information concerning school law, which has never before been accessible to either teacher or school officer.

A. ARMSTRONG,
Supt. Schools, Sioux City, Iowa.

DES MOINES, May 15, 1880.

The examination of "A Treatise on the Law of Public Schools," prepared by Finley Burke, Esq., of Council Bluffs, has given me much pleasure. So far as I know, there is no work of similar character now in existence. I think such a work will be exceedingly useful to lawyers, school officers, and teachers, and I hope that it may find its way into their hands.

G. W. VON COELLN,
Supt. Public Inst. for Iowa.

Teachers' Handbook. Phelps.

By William F. Phelps, Principal of Minnesota State Normal School. Embracing the objects, history, organization, and management of teachers' institutes, followed by methods of teaching, in detail, for all the fundamental branches. Every young teacher, every practical teacher, every experienced teacher even, needs this book.

From the New York Tribune.

"The discipline of the school should prepare the child for the discipline of life. The country schoolmaster, accordingly, holds a position of vital interest to the destiny of the republic, and should neglect no means for the wise and efficient discharge of his significant functions. This is the key-note of the present excellent volume. In view of the supreme importance of the teacher's calling, Mr. Phelps has presented an elaborate system of instruction in the elements of learning, with a complete detail of methods and processes, illustrated with an abundance of practical examples and enforced by judicious councils."

Topical Course of Study. Stone.

This volume is a compilation from the courses of study of our most successful public schools, and the best thought of leading educators. The pupil is enabled to make full use of any and all text-books bearing on the given topics, and is incited to use all other information within his reach.

American Education. Mansfield.

A treatise on the principles and elements of education, as practised in this country, with ideas towards distinctive republican and Christian education.

American Institutions. De Tocqueville.

A valuable index to the genius of our Government.

Universal Education. Mayhew.

The subject is approached with the clear, keen perception of one who has observed its necessity, and realized its feasibility and expediency alike. The redeeming and elevating power of improved common schools constitutes the inspiration of the volume.

Oral Training Lessons. Barnard.

The object of this very useful work is to furnish material for instructors to impart orally to their classes, in branches not usually taught in common schools, embracing a' departments of natural science and much general knowledge.

Lectures on Natural History. Chadbourne.

Affording many themes for oral instruction in this interesting science, especially in schools where it is not pursued as a class exercise.

MISCELLANEOUS PUBLICATIONS—*Continued.*

Outlines of Mathematical Science. Davies.

A manual suggesting the best methods of presenting mathematical instruction on the part of the teacher, with that comprehensive view of the whole which is necessary to the intelligent treatment of a part, in science.

Nature and Utility of Mathematics. Davies.

An elaborate and lucid exposition of the principles which lie at the foundation of pure mathematics, with a highly ingenious application of their results to the development of the essential idea of the different branches of the science.

Mathematical Dictionary. Davies and Peck.

This cyclopædia of mathematical science defines, with completeness, precision, and accuracy, every technical term; thus constituting a popular treatise on each branch, and a general view of the whole subject.

The Popular Educator. Barnes.

In seven volumes, containing interesting and profitable educational miscellany.

Liberal Education of Women. Orton.

Treats of "the demand and the method;" being a compilation of the best and most advanced thought on this subject, by the leading writers and educators in England and America. Edited by a professor in Vassar College.

Education Abroad. Northrop.

A thorough discussion of the advantages and disadvantages of sending American children to Europe to be educated; also, papers on legal prevention of illiteracy, study, and health, labor as an educator, and other kindred subjects.

The Teacher and the Parent. Northend.

A treatise upon common-school education, designed to lead teachers to view their calling in its true light, and to stimulate them to fidelity.

The Teachers' Assistant. Northend.

A natural continuation of the author's previous work, more directly calculated for daily use in the administration of school discipline and instruction.

School Government. Jewell.

Full of advanced ideas on the subject which its title indicates. The criticisms upon current theories of punishment and schemes of administration have excited general attention and comment.

Grammatical Diagrams. Jewell.

The diagram system of teaching grammar explained, defended, and improved. The curious in literature, the searcher for truth, those interested in new inventions, as well as the disciples of Professor Clark, who would see their favorite theory fairly treated, all want this book. There are many who would like to be made familiar with this system before risking its use in a class. The opportunity is here afforded.

The Complete Examiner. Stone.

Consists of a series of questions on every English branch of school and academic instruction, with reference to a given page or article of leading text-books where the answer may be found in full. Prepared to aid teachers in securing certificates, pupils in preparing for promotion, and teachers in selecting review questions.

How Not to Teach. Griffin.

This book meets a want universally felt among young teachers who have their experience in teaching to learn. It undertakes to point out the many natural mistakes into which teachers, unconsciously or otherwise, fall, and warns the reader against dangers that beset the path of every conscientious teacher. It tells the reader, also, the proper and acceptable way to teach, illustrating the author's ideas by some practice-lessons in arithmetic (after Grübe).

MISCELLANEOUS PUBLICATIONS — *Continued.*

School Amusements. Root.

To assist teachers in making the school interesting, with hints upon the management of the school-room. Rules for military and gymnastic exercises are included. Illustrated by diagrams.

Institute Lectures. Bates.

These lectures, originally delivered before institutes, are based upon various topics in the departments of mental and moral culture. The volume is calculated to prepare the will, awaken the inquiry, and stimulate the thought of the zealous teacher.

Method of Teachers' Institutes. Bates.

Sets forth the best method of conducting institutes, with a detailed account of the object, organization, plan of instruction, and true theory of education on which such instruction should be based.

History and Progress of Education.

The systems of education prevailing in all nations and ages, the gradual advance to the present time, and the bearing of the past upon the present, in this regard, are worthy of the careful investigation of all concerned in education.

Higher Education. Atlas Series.

A collection of valuable essays. CONTENTS. International Communication by Language, by Philip Gilbert Hamerton; Reform in Higher Education; Upper Schools, by President James McCosh; Study of Greek and Latin Classics, by Prof. Charles Elliott; The University System in Italy, by Prof. Angelo de Gubernatis, of the University of Florence; Universal Education, by Ray Palmer; Industrial Art Education, by Eaton S. Drone.

LIBRARY OF LITERATURE.

Milton's Paradise Lost. (Boyd's Illustrated Edition.)
Young's Night Thoughts. do.
Cowper's Task, Table Talk, &c. do.
Thomson's Seasons. do.
Pollok's Course of Time. do.

These works, models of the best and purest literature, are beautifully illustrated, and notes explain all doubtful meanings.

Lord Bacon's Essays. (Boyd's Edition.)

Another grand English classic, affording the highest example of purity in language and style.

The Iliad of Homer. (Translated by Pope.)

Those who are unable to read this greatest of ancient writers in the original should not fail to avail themselves of this standard metrical version.

Pope's Essay on Man.

This is a model of pure classical English, which should be read, also, by every teacher and scholar for the sound thought it contains.

Improvement of the Mind. Isaac Watts.

No mental philosophy was ever written which is so comprehensive and practically useful to the unlearned as well as learned reader as this well-known book of Watts.

Milton's Political Works. Cleveland.

This is the very best edition of the great poet. It includes a life of the author, notes, dissertations on each poem, a faultless text, and is *the only* edition of Milton with a complete verbal index.

www.ingramcontent.com/pod-product-compliance
Lightning Source LLC
LaVergne TN
LVHW050519100826
845148LV00002B/386

* 9 7 8 1 4 2 5 5 2 0 7 7 9 *